U.S. Department of Labor
Occupational Safety
and Health Administration

Office of General Industry
Compliance Assistance

OSHA
Field Inspection
Reference Manual

Government Institutes, Inc.

Government Institutes, Inc., 4 Research Place, Suite 200, Rockville, Maryland 20850

99   98   97   96   5   4   3

This publication was prepared by the U.S. Department of Labor, Occupational Safety and Health Administration, for use within OSHA. Government Institutes determined that this material was of interest to those outside of OSHA and reproduced it in order to serve those interested.

The OSHA Field Inspection Reference Manual (FIRM) was developed by the Office of General Industry Compliance Assistance for compliance officers and is dated September 1994. This instruction (CPL 2.103) is designed to provide the field offices with a reference document for identifying the responsibilities associated with the majority of their inspection duties.

The issuance of the FIRM will require a change to the OSHA Field Operations Manual (FOM), but until this change is accomplished, if there are any discrepancies between the FIRM and the FOM, the FIRM prevails.

ISBN: 0-86587-426-3

Printed in the United States of America

U.S. Department of Labor   Assistant Secretary for
Occupational Safety and Health
Washington, D.C. 20210

OSHA Instruction CPL 2.103
SEP 2 6 1994
Office of General Industry Compliance Assistance

SUBJECT: Field Inspection Reference Manual (FIRM)

A.   Purpose. The purpose of this instruction is to transmit the Field Inspection Reference Manual (FIRM). The FIRM was developed by the Field Operations Manual (FOM) Revision Team to provide the field offices a reference document for identifying the responsibilities associated with the majority of their inspection duties.

B.   Scope. This instruction applies OSHA-wide.

C.   References.

1.   OSHA Instruction CPL 2.45B, June 15, 1989, Field Operations Manual (FOM).

2.   OSHA Instruction CPL 2.51H, March 22, 1993, Exemptions and Limitations under the Current Appropriations Act.

3.   OSHA Instruction CPL 2.77, December 30, 1986, Critical Fatality/Catastrophe Investigation.

4.   OSHA Instruction CPL 2.80, October 21, 1990, Handling of Cases to be Proposed for Violation-by-Violation Penalties.

5.   OSHA Instruction CPL 2.90, June 3, 1991, Guidelines for Administration of Corporate-Wide Settlement Agreements.

6.   OSHA Instruction CPL 2.94, July 22, 1991, OSHA Response to Significant Events of Potentially Catastrophic Consequences.

7.   OSHA Instruction CPL 2.97, January 26, 1993, Fatality/Catastrophe Reports to the National Office ("Flash Reports").

8.   OSHA Instruction CPL 2.98, October 12, 1993, Guidelines for Case File Documentation for Use with Videotapes and Audiotapes.

9.   OSHA Instruction CPL 2.102, March 28, 1994, Procedures for Approval of Local Emphasis Programs (LEPs) and Experimental Programs.

10. OSHA Instruction CPL 2-2.20B, April 19, 1993, OSHA Technical Manual.

11. OSHA Instruction CPL 2-2.35A, December 19, 1983, 29 CFR 1910.95(b)(1), Guidelines for Noise Enforcement, Appendix A, Noise Control Guidelines.

12. OSHA Instruction CPL 2-2.54, February 10, 1992, Respiratory Protection Program Manual.

13. OSHA Instruction ADM 1-1.31, September 20, 1993, The IMIS Enforcement Data Processing Manual.

14. OSHA Instruction ADM 4.4, August 19, 1991, Administrative Subpoenas.

15. OSHA Instruction ADM 11.3A, July 29, 1991, Revision A to the OSHA Mission and Function Statements.

16. Memorandum dated March 31, 1994, to the Regional Administrators from H. Berrien Zettler, Deputy Director, Directorate of Compliance Programs regarding Policy Regarding Voluntary Rescue Activities.

D. Action. The issuance of the FIRM will require a change to the FOM, but until this change is accomplished, if there are any discrepancies between the FIRM and the FOM, the FIRM prevails. Regional Administrators and Area Directors shall ensure that the policies and procedures established in this instruction are transmitted to all Area and District Offices, and to appropriate staff.

E. Effective Date. September 30, 1994.

F. Federal Agencies. This instruction describes a change that affects Federal agencies. Executive Order 12196, Section 1-201, and 29 CFR 1960.16, maintains that Federal agencies must also follow the enforcement policy and procedures contained in this instruction.

G. Federal Program Change. This instruction describes a Federal program change which affects State programs. Each Regional Administrator shall:

1. Ensure that this change is promptly forwarded to each State designee using a format consistent with the Plan Change Two-way Memorandum in Appendix P, OSHA Instruction STP 2.22A, CH-3. Explain the content of this change to the State designee as requested.

2. Encourage the State designees to review the streamlined procedures contained in the Field Inspection Reference Manual (FIRM) and adopt parallel procedures through a State FIRM or changes to the State Field Operations Manual (FOM).

3. Advise the State designees that specific policy changes have been made in the following areas, through the FIRM, which require a State response:

    a. serious willful penalty increase (previously transmitted by June 14, 1994, memorandum "Revised Penalties and Willful Violations") - FIRM Chapter IV, C.2.m.
    b. good faith credit - FIRM Chapter IV, C.2.i.(5)(b)
    c. definitions concerning complaints - FIRM Chapter I, C.2.
    d. nonformal complaint procedures - FIRM Chapter I, C.7.
    e. voluntary rescue operations - FIRM Chapter II, B.2.e.
    f. imminent danger investigations - FIRM Chapter II, B.3.
    g. unobserved exposure - FIRM Chapter III, C.1.b.(4)
    h. reporting of fatalities/catastrophe (1904.8) - FIRM Chapter II, B.2.a.
    i. reporting time for fatalities (1904.8) - FIRM Chapter IV, C.2.n.(3)(c)
    j. notification of employee representatives of citation issuance - FIRM Chapter IV, B.1.b.
    k. economic feasibility - FIRM Chapter IV, A.6.a.(4)(b) and NOTE
    l. employee involvement in informal conferences/settlements - FIRM Chapter IV, D.1.b.

4. Ensure that the State designees are asked to acknowledge receipt of this Federal program change in writing to the Regional Administrator as soon as possible, but not later than 70 calendar days after the date of issuance (10 days for mailing and 60 days for response). The acknowledgment must include a statement indicating the State's general intention with regard to adopting the FIRM or incorporating the FIRM's procedures into its Field Manual and specific intention in regard to the 12 policy changes specified above.

5. Ensure that State designees submit an appropriate State plan supplement within 6 months and that it is reviewed and processed in accordance with paragraphs I.1.a.(3)(a) and (b), Part I of the State Plan Policies and Procedures Manual (SPM).

6.   Review policies, instructions and guidelines issued by the State to determine that this change has been communicated to State compliance personnel.

Joseph A. Dear
Joseph A. Dear
Assistant Secretary

Ron Yarman
Ron Yarman
Executive Vice President, NCFLL

DISTRIBUTION: National, Regional, and Area Offices
All Compliance Officers
State Designees
NIOSH Regional Program Directors
7(c)(1) Project Managers

TABLE OF CONTENTS

PAGE

CHAPTER II. INSPECTIONS PROCEDURES

A. GENERAL INSPECTION PROCEDURES

CHAPTER III   INSPECTION DOCUMENTATION

This manual is intended to provide guidance regarding some of the internal operations of the Occupational Safety and Health Administration (OSHA), and is solely for the benefit of the Government. No duties, rights, or benefits, substantive or procedural, are created or implied by this manual. The contents of this manual are not enforceable by any person or entity against the Department of Labor or the United States. Guidelines which reflect current Occupational Safety and Health Review Commission or court precedents do not necessarily indicate acquiescence with those precedents.

CHAPTER I

PRE-INSPECTION PROCEDURES

A. <u>General Responsibilities and Administrative Procedures</u>. The following are brief descriptions of the general responsibilities for positions within OSHA. Reference OSHA Instruction ADM 11.3A for complete information on OSHA mission and function statements. Employees should refer to their position descriptions for individual job responsibilities. This document empowers OSHA personnel to make decisions as situations warrant with the ability to act efficiently to accomplish the mission of OSHA and to enforce the Occupational Safety and Health Act.

1. <u>Regional Administrator</u>. It is the duty or mission of the Regional Administrator to manage, execute and evaluate all programs of the Occupational Safety and Health Administration (OSHA) in the region. The Regional Administrator reports to the Assistant Secretary through the career Deputy Assistant Secretary.

2. <u>Area Director (AD)</u>. It is the duty or mission of the Area Director to accomplish OSHA's programs within the delegated area of responsibility of the Area Office. This includes administrative and technical support of the Compliance Safety and Health Officers (CSHOs) assigned to the Area Office, and the issuing of citations.

3. <u>Assistant Area Director (AAD)</u>. The Assistant Area Director has first level supervisory responsibility over CSHOs in the discharge of their duties and may also conduct compliance inspections. Assistant Area Directors ensure technical adequacy in applying the policies and procedures in effect in the Agency. The Assistant Area Director shall implement a quality assurance system suitable to the work group, including techniques such as random review of selected files, review based on CSHO recommendation/request, and verbal briefing by CSHOs and/or review of higher profile or non-routine cases.

4. <u>Compliance Safety and Health Officer</u>.

   a. <u>General</u>. The primary responsibility of the Compliance Safety and Health Officer (CSHO) is to carry out the mandate given to the Secretary of Labor, namely, "to assure so far as possible every working man and woman in the Nation safe and healthful working conditions...." To accomplish this mandate the Occupational Safety and Health Administration employs a wide variety of programs and initiatives, one of which is enforcement of standards through the conduct of effective inspections to determine whether employers are:

      (1) Furnishing places of employment free from recognized hazards that are causing or are likely to cause death or serious physical harm to their employees, and

(2) Complying with safety and health standards and regulations.

Through inspections and other employee/employer contact, the CSHO can help ensure that hazards are identified and abated to protect workers. During these processes, the CSHO must use professional judgment to adequately document hazards in the case file, as required by the policies and procedures in effect in the Agency. The CSHO will be responsible for the technical adequacy of each case file.

A. 4.   b.   Subpoena Served on CSHO. If a CSHO is served with a subpoena, the Area Director shall be informed immediately and shall refer the matter to the Regional Solicitor.

c.   Testifying in Hearings. The CSHO is required to testify in hearings on OSHA's behalf. The CSHO shall be mindful of this fact when recording observations during inspections. The case file shall reflect conditions observed in the workplace as accurately as possible. If the CSHO is called upon to testify, the case file will be invaluable as a means for recalling actual conditions.

d.   Release of Inspection Information. The information obtained during inspections is confidential, but is to be determined as disclosable or nondisclosable on the basis of criteria established in the Freedom of Information Act, as amended, in 29 CFR Part 70, and in Chapter XIV of OSHA Instruction CPL 2.45B or a superseding directive. Requests for release of inspection information shall be directed to the Area Director.

e.   Disposition of Inspection Records. "Inspection Records" are any records made by a CSHO that concern, relate to, or are part of any inspection or that concern, relate to, or are part of the performance of any official duty. Such original material and all copies shall be included in the case file. These records are the property of the United States Government and a part of the case file. Inspection records are not the property of the CSHO and under no circumstances are they to be retained or used for any private purpose. Copies of documents, notes or other recorded information not necessary or pertinent, or not suitable for inclusion in the case file shall, with the concurrence and permission of the Area Director, be destroyed in accordance with an approved record disposition schedule.

5.   General Area Office Responsibilities. The Area Director shall ensure that the Area Office maintains an outreach program appropriate to local conditions and the needs of the service area. The plan may include utilization of Regional Office Support Services, training and education services, referral services, voluntary compliance programs, abatement assistance, and technical services.

B. Inspection Scheduling.

1. Program Planning. Effective and efficient use of resources requires careful, flexible planning. In this way, the overall goal of hazard abatement and worker protection is best served.

2. Inspection/Investigation Types.

   a.   Unprogrammed. Inspections scheduled in response to alleged hazardous working conditions that have been identified at a specific worksite are unprogrammed. This type of inspection responds to reports of imminent dangers, fatalities/catastrophes, complaints and referrals. It also includes followup and monitoring inspections scheduled by the Area Office.

        NOTE: This category includes all employers directly affected by the subject of the unprogrammed activity, and is especially applicable on multiemployer worksites.

   b.   Unprogrammed Related. Inspections of employers at multiemployer worksites whose operations are not directly affected by the subject of the conditions identified in the complaint, accident, or referral are unprogrammed related. An example would be a trenching inspection conducted at the unprogrammed worksite, where the trenching hazard was not identified in the complaint, accident report, or referral.

   c.   Programmed. Inspections of worksites which have been scheduled based upon objective or neutral selection criteria are programmed. The worksites are selected according to national scheduling plans for safety and for health or special emphasis programs.

   d.   Programmed Related. Inspections of employers at multiemployer worksites whose activities were not included in the programmed assignment, such as a low injury rate employer at a worksite where programmed inspections are being conducted for all high injury rate employers. All high hazard employers at the worksite shall normally be included in the programmed inspections. (See Chapter II, F.2. of OSHA Instruction CPL 2.45B or a superseding directive).

B. 3. <u>Inspection Priorities</u>.

    a.   <u>Order of Priority</u>. Generally, priority of accomplishment and assignment of staff resources for inspection categories shall be as follows:

Priority	Category
First	Imminent Danger
Second	Fatality/Catastrophe Investigations
Third	Complaints/Referrals Investigations
Fourth	Programmed Inspections

    b.   <u>Efficient Use of Resources</u>. Based on the nature of the alleged hazard, unprogrammed inspections normally shall be scheduled and conducted prior to programmed inspections. Deviations from this priority list are allowed so long as they are justifiable, lead to efficient use of resources, **and** contribute to the effective protection of workers. An example of such a deviation would be for Area Directors to commit a certain percentage of IH resources to programmed Local Emphasis Program (LEP) inspections.

    c.   <u>Followup Inspections</u>. In cases where followup inspections are necessary, they shall be conducted as promptly as resources permit. Except in unusual circumstances, followup inspections shall take priority over all programmed inspections and any unprogrammed inspections with hazards evaluated as other than serious. Followup inspections should not normally be conducted within the 15 working day contest period unless high gravity serious violations were issued. See Chapter IV at A.4. regarding effect of contest upon abatement period.

4. <u>Inspection Selection Criteria</u>.

    a.   <u>General Requirements</u>. OSHA's priority system for conducting inspections is designed to distribute available OSHA resources as effectively as possible to ensure that maximum feasible protection is provided to the working men and women of this country.

        (1)   <u>Scheduling</u>. The Area Director shall ensure that inspections are scheduled within the framework of this chapter, that they are consistent with the objectives of the Agency, and that appropriate documentation of scheduling practices is maintained. (See OSHA Instruction CPL 2.51H, or most current version, for congressional exemptions and limitations on OSHA inspection activity.)

B. 4. a.　(2)　<u>Effective Use of Resources</u>. The Area Director shall ensure that OSHA resources are effectively distributed during inspection activities. If an inspection is of a complex nature, the Area Director may consider utilizing outside OSHA resources (e.g., the Health Response Team) to more effectively employ the Area or District Office resources. The Area Office will retain control of the inspection.

(3)　<u>Effect of Contest</u>. If an employer scheduled for inspection, either programmed or unprogrammed, has contested a citation and/or a penalty received as a result of a previous inspection and the case is still pending before the Review Commission, the following guidelines apply:

(a)　If the employer has contested the penalty only, the inspection shall be scheduled as though there were no contest.

(b)　If the employer has contested the citation itself or any items thereon, then programmed and unprogrammed inspections shall be scheduled in accordance with the guidelines in B.3.a. of this chapter. The scope of unprogrammed inspections normally shall be partial. All items under contest shall be excluded from the inspection unless a potential imminent danger is involved.

b.　<u>Employer Contacts</u>. Contacts for information initiated by employers or their representatives shall not trigger an inspection, nor shall such employer inquiries protect them against regular inspections conducted pursuant to guidelines established by the agency. Further, if an employer or its representative indicates that an imminent danger exists or that a fatality or catastrophe has occurred, the Area Director shall act in accordance with established inspection priority procedures.

C.　<u>Complaints & Other Unprogrammed Inspections</u>.

1.　<u>General</u>. This section relates to information received and processed at the area office before an inspection rather than information which is given to the CSHO during an inspection. Complaints will be evaluated according to local Area Office procedures, including using the criteria established in Chapter III, C.2. for classifying the alleged violations as serious or other. When essential information is not provided by the complainant, the complaint is too vague to evaluate, or the Area Office has other specific information that the complaint is not valid, an attempt shall be made to clarify or supplement available information. If a decision is made that the complaint is not valid, a letter will be sent to the complainant advising of the decision and its reasons.

C. 2. <u>Definitions</u>.

  a. <u>Complaint (OSHA-7)</u>. Notice of an alleged hazard (over which OSHA has jurisdiction), or a violation of the Act, given by a past or present employee, a representative of employees, a concerned citizen, or an 11(c) officer seeking resolution of a discrimination complaint.

  b. <u>Referral (OSHA-90)</u>. Notice of an alleged hazard or a violation of the Act given by any source not listed under C.2.a., above, including media reports.

  c. <u>Formal Complaint</u>. A signed complaint alleging an imminent danger or the existence of a violation threatening physical harm, submitted by a current employee, a representative of employees (such as unions, attorneys, elected representatives, and family members), or present employee of another company if that employee is exposed to the hazards of the complained-about workplace. Reference Section 8(f) of the Act and 29 CFR §1903.11.

  d. <u>Nonformal Complaint</u>. Oral or unsigned complaints, or complaints by former employees or non-employees.

3. <u>Identity of Complainant</u>. The identity of the complainant shall be kept confidential unless otherwise requested by the complainant, in accordance with Section 8(f)(1) of the Act. No information shall be given to employers which would allow them to identify the complainant.

4. <u>Formalizing Oral Complaints</u>.

  a. If the complainant meets the criteria in C.2.c., above, for filing formal complaints and wishes to formalize an oral complaint, all pertinent information will be entered on an OSHA-7 form, or equivalent, by the complainant or a member of the Area Office staff. A copy of this completed form can be sent to the complainant for signature, or the complainant shall be asked to sign a letter with the particular details of the complaint to the area office.

  b. The complainant shall be informed that, if the signed complaint is not returned within 10 working days, it shall be treated as a nonformal complaint. If the signed complaint arrives after the 10 working days but prior to OSHA's contact with the employer, it will be treated as a formal complaint.

  c. The following are examples of deficiencies which would result in the failure of an apparent formal complaint to meet the requirements of the definition:

(1)   A thorough evaluation of the complaint does not establish reasonable grounds to believe that the alleged violation can be classified as an imminent danger or that the alleged hazard is covered by a standard or, in the case of an alleged serious condition, by the general duty clause (Section 5(a)(1)).

(2)   The complaint concerns a workplace condition which has no direct relationship to safety or health and does not threaten physical harm; e.g., a violation of a regulation or a violation of a standard that is classified as de minimis.

(3)   The complaint alleges a hazard which violates a standard but describes no actual workplace conditions and gives no particulars which would allow a proper evaluation of the hazard. In such a case the Area Director shall make a reasonable attempt to obtain such information.

C. 5.   <u>Imminent Danger Report Received By the Field</u>. Any allegation of imminent danger received by an OSHA office shall be handled in accordance with the following procedures:

a.   The Area Director shall immediately determine whether there is a reasonable basis for the allegation.

b.   Imminent danger investigations shall be scheduled with the highest priority.

c.   When an immediate inspection cannot be made, the Area Director or CSHO shall contact the employer (and when known, the employee representative) immediately, obtain as many pertinent details as possible concerning the situation and attempt to have any employees affected by imminent danger voluntarily removed. Such notification shall be considered advance notice and shall be handled in accordance with the procedures given in E.3. of this chapter.

6.   <u>Formal Complaints</u>. All formal complaints meeting the requirements of Section 8(f)(1) of the Act and 29 CFR 1903.11 shall be scheduled for workplace inspections.

a.   <u>Determination</u>. Upon determination by the Assistant Area Director that a complaint is formal, an inspection shall be scheduled in accordance with the priorities in C.6.b.

b.   <u>Priorities for Responding by Inspections to Formal Complaints</u>. Inspections resulting from formal complaints shall be conducted according to the following priority:

(1)　Formal complaints, other than imminent danger, shall be given a priority based upon the classification and the gravity of the alleged hazards as defined in Chapters III and IV.

(2)　Formal serious complaints shall be investigated on a priority basis within 30 working days and formal other-than-serious complaints within 120 days.

(3)　If resources do not permit investigations within the time frames given in (2), a letter to the complainant shall explain the delay and shall indicate when an investigation may occur. The complainant shall be asked to confirm the continuation of the alleged hazardous conditions.

(4)　If a late complaint inspection is to be conducted, the Area Director may contact the complainant to ensure that the alleged hazards are still existent.

C. 7.　<u>Nonformal Complaints</u>.

a.　<u>Serious</u>.　If a decision is made to handle a serious nonformal complaint by letter, a certified letter shall be sent to the employer advising the employer of the complaint items and the need to respond to OSHA within a specified time. When applicable, the employer shall be informed of Section 11(c) requirements, and that the complainant will be kept informed of the complaint progress. Follow-up contact may be by telephone at the option of the Area Office.

b.　<u>Other-Than-Serious and De Minimis</u>.　Nonformal complaints about other-than-serious hazards or de minimis conditions may be investigated by telephone if they can be satisfactorily resolved in that manner, with follow-up telephone contact to the complainant with the results of the employer's investigation and corrective actions. Corrective action shall be documented in the case file. If, however, the telephone contact is inadequate, a letter will be sent to the employer.

c.　<u>Letter to Complainant</u>.　Concurrent with the letter to the employer, a letter to the complainant shall be sent containing a copy of the letter to the employer. The complainant will be asked to notify OSHA if no corrective action is taken within the indicated time frame, or if any adverse or discriminatory action or threats are made due to the complainant's safety and health activities. Copies of subsequent correspondence related to the complaint shall be sent to the complainant, if requested.

d.   Inspection of Nonformal Complaint.

(1)   Where the employer fails to respond or submits an inadequate response, the employer may be contacted to find out what corrections will be made, or the nonformal complaint will be activated for inspection. If no action has been taken, the nonformal complaint shall normally be activated for inspection.

(2)   Nonformal complaints, when received by the Area Office, may be activated for inspection if the Area Director or representative judges the hazard to be high gravity serious in nature, and the inspection can be performed with efficient use of resources.

C. 8.   Results of Inspection to Complainant.   After an inspection, the complainant shall be sent a letter addressing each complaint item, with reference to the citations and/or with a sufficiently detailed description of the findings and why they did not result in a violation. The complainant shall also be informed of the appeal rights under 29 CFR 1903.12.

9.   Discrimination Complaints.   The complainant shall be advised of the protection against discrimination afforded by Section 11(c) of the Act and shall be informed of the procedure for filing an 11(c) complaint.

a.   Safety and/or health complaints filed by former employees who allege that they were fired for exercising their rights under the Act are nonformal complaints and will not be scheduled for investigation.

(1)   Such complaints shall be recorded on an OSHA-7 Form and handled in accordance with the procedures outlined in OSHA Instruction DIS .4B. They shall be transmitted to the appropriate 11(c) personnel for investigation of the alleged 11(c) discrimination complaint.

(2)   No letter shall be sent to the employer until after the Regional Supervisory Investigator has reviewed the case and decided that no recommendation for inspection will be submitted to the Area Director.

(3)   This screening process by the Regional Supervisory Investigator is not anticipated to take more than 3 work days and usually less. The Area Director can expect to be informed by telephone of the decision within that time frame.

b.   In those instances where the Regional Supervisory Investigator determines that the existence or nature of the alleged hazard is likely to be relevant to the resolution of the 11(c) discrimination complaint, the complaint shall be sent back to the Area Director for an OSHA inspection to be handled as a referral.

c.   When the decision is that no inspection is necessary, the Area Director shall ensure that the complaint has been recorded on an OSHA-7 Form and proceed to send a letter to the employer in accordance with procedures for responding to nonformal complaints.

d.   Any 11(c) complaint alleging an imminent danger shall be handled in accordance with the instructions in C.5.

C. 10. Referrals. Referrals shall be handled in a manner similar to that of complaints.

a.   Letters. Referrals shall normally be handled by letter or telephone. For those referrals handled by letter, complaint letters can be revised to fit the particular circumstances of the referral.

b.   Inspections. High gravity serious referrals shall normally be handled by inspection. A letter transmitting the results of the investigation shall be sent to any referring agency/department.

11. Accidents. Accidents involving significant publicity or any other accident not involving a fatality or a catastrophe, however reported, may be considered as either a complaint or a referral, depending on the source of the report, and shall be handled according to the directions given in Chapter II.

D. Programmed Inspections.

1.   Programmed inspections shall be scheduled in accordance with Chapter II, F.2. of OSHA Instruction CPL 2.45B or a superseding directive.

2.   Local emphasis programmed inspections shall be conducted in accordance with OSHA Instruction CPL 2.102.

E. Inspection Preparation.

1.   General. The conduct of effective inspections requires professional judgment in the identification, evaluation and reporting of safety and health conditions and practices. Inspections may vary considerably in scope and detail, depending upon the circumstances in each case.

2.   Planning. It is most important that the CSHO adequately prepares for an inspection. The CSHO shall also ensure the selection of appropriate inspection materials and equipment, including personal protective equipment, based on anticipated

exposures and training received in relation to the uses and limitations of such equipment. Refer to OSHA Instruction CPL 2-2.54 regarding respiratory protection.

a. 29 CFR §1903.7(c) requires that the CSHO comply with all safety and health rules and practices at the establishment and wear or use the safety clothing or protective equipment required by OSHA standards or by the employer for the protection of employees.

b. The CSHO shall not enter any area where special entrance restrictions apply until the required precautions have been taken. It shall be the Assistant Area Director's responsibility to procure whatever materials and equipment are needed for the safe conduct of the inspection.

E. 3. <u>Advance Notice of Inspections</u>.

a. <u>Policy</u>. Section 17(f) of the Act and 29 CFR §1903.6 contain a general prohibition against the giving of advance notice of inspections, except as authorized by the Secretary or the Secretary's designee.

(1) The Occupational Safety and Health Act regulates many conditions which are subject to speedy alteration and disguise by employers. To forestall such changes in worksite conditions, the Act, in Section 8(a), prohibits unauthorized advance notice.

(2) There may be occasions when advance notice is necessary to conduct an effective investigation. These occasions are narrow exceptions to the statutory prohibition against advance notice.

(3) Advance notice of inspections may be given only with the authorization of the Area Director and only in the following situations:

(a) In cases of apparent imminent danger to enable the employer to correct the danger as quickly as possible;

(b) When the inspection can most effectively be conducted after regular business hours or when special preparations are necessary;

(c) To ensure the presence of employer and employee representatives or other appropriate personnel who are needed to aid in the inspection; and

(d) When the giving of advance notice would enhance the probability of an effective and thorough inspection; e.g., in complex fatality investigations.

(4)   Advance notice exists whenever the Area Office sets up a specific date or time with the employer for the CSHO to begin an inspection. Any delays in the conduct of the inspection shall be kept to an absolute minimum. Lengthy or unreasonable delays shall be brought to the attention of the Assistant Area Director. Advance notice generally does not include nonspecific indications of potential future inspections.

(5)   In unusual circumstances, the Area Director may decide that a delay is necessary. In those cases the employer or the CSHO shall notify affected employee representatives, if any, of the delay and shall keep them informed of the status of the inspection.

b.   <u>Documentation</u>. The conditions requiring advance notice and the procedures followed shall be documented in the case file.

E. 4.   <u>Pre-Inspection Compulsory Process</u>. 29 CFR § 1903.4 authorizes the agency to seek a warrant in advance of an attempted inspection if circumstances are such that "pre-inspection process (is) desirable or necessary." The Act authorizes the agency to issue administrative subpoenas to obtain relevant information.

a.   Although the agency generally does not seek warrants without evidence that the employer is likely to refuse entry, the Area Director may seek compulsory process in advance of an attempt to inspect or investigate whenever circumstances indicate the desirability of such warrants.

NOTE: Examples of such circumstances would be evidence of denied entry in previous inspections, or awareness that a job will only last a short time or that job processes will be changing rapidly.

b.   Administrative subpoenas may also be issued prior to any attempt to contact the employer or other person for evidence related to an OSHA inspection or investigation. (See OSHA Instruction ADM 4.4. and Chapter II, A.2.c.(3).)

5.   <u>Expert Assistance</u>.

a.   The Area Director shall arrange for a specialist and/or specialized training, preferably from within OSHA, to assist in an inspection or investigation when the need for such expertise is identified.

b.   OSHA specialists may accompany the CSHO or perform their tasks separately. A CSHO must accompany outside consultants. OSHA specialists and outside consultants shall be briefed on the purpose of the inspection and personal protective equipment to be utilized.

E. 6. Personal Security Clearance. Some establishments have areas which contain material or processes which are classified by the U.S. Government in the interest of national security. Whenever an inspection is scheduled for an establishment containing classified areas, the Assistant Area Director shall assign a CSHO who has the appropriate security clearances. The Regional Administrator shall ensure that an adequate number of CSHOs with appropriate security clearances are available within the Region and that the security clearances are current.

7. Disclosure of Records. The disclosure of inspection records is governed by the Department's regulations at 29 CFR part 70, implementing the Freedom of Information Act (FOIA).

8. Classified and Trade Secret Information. Any classified or trade secret information and/or personal knowledge of such information by OSHA personnel shall be handled in accordance with the regulations of OSHA or of the responsible agency. The collection of such information, and the number of personnel with access to it shall be limited to the minimum necessary for the conduct of compliance activities. The CSHO shall identify classified and trade secret information as such in the case file. Title 18 USC, Section 1905, as referenced by Section 15 of the OSH Act, provides for criminal penalties in the event of improper disclosure.

CHAPTER II

INSPECTION PROCEDURES

A. <u>General Inspection Procedures</u>.

1. <u>Inspection Scope</u>. Inspections, either programmed or unprogrammed, fall into one of two categories depending on the scope of the inspection:

   a. <u>Comprehensive</u>. A substantially complete inspection of the potentially high hazard areas of the establishment. An inspection may be deemed comprehensive even though, as a result of the exercise of professional judgment, not all potentially hazardous conditions, operations and practices within those areas are inspected.

   b. <u>Partial</u>. An inspection whose focus is limited to certain potentially hazardous areas, operations, conditions or practices at the establishment. A partial inspection may be expanded based on information gathered by the CSHO during the inspection process. Consistent with the provisions of Section 8(f)(2) of the Act, and Area Office priorities, the CSHO shall use professional judgment to determine the necessity for expansion of the inspection scope, based on information gathered during records or program review and walka-round inspection.

2. <u>Conduct of the Inspection</u>.

   a. <u>Time of Inspection</u>. Inspections shall be made during regular working hours of the establishment except when special circumstances indicate otherwise. The Assistant Area Director and CSHO shall confer with regard to entry during other than normal working hours.

   b. <u>Presenting Credentials</u>.

      (1) At the beginning of the inspection the CSHO shall locate the owner representative, operator or agent in charge at the workplace and present credentials. On construction sites this will most often be the representative of the general contractor.

      (2) When neither the person in charge nor a management official is present, contact may be made with the employer to request the presence of the owner, operator or management official. The inspection shall not be delayed unreasonably to await the arrival of the employer representative. This delay should normally not exceed one hour. If the person in charge at the workplace cannot be determined, record the extent of the inquiry in the case file and proceed with the physical inspection.

A. 2. c.   <u>Refusal to Permit Inspection</u>. Section 8 of the Act "provides that CSHOs may enter without delay and at reasonable times any establishment covered under the Act for the purpose of conducting an inspection". Unless the circumstances constitute a recognized exception to the warrant requirement (i.e., consent, third party consent, plain view, open field, or exigent circumstances) an employer has a right to require that the CSHO seek an inspection warrant prior to entering an establishment and may refuse entry without such a warrant.

> **NOTE:**   On a military base or other Federal Government facility, the following guidelines do not apply. Instead, a representative of the controlling authority shall be informed of the contractor's refusal and asked to take appropriate action to obtain cooperation.

(1)   <u>Refusal of Entry or Inspection</u>. When the employer refuses to permit entry upon being presented proper credentials or allows entry but then refuses to permit or hinders the inspection in some way, a tactful attempt shall be made to obtain as much information as possible about the establishment. (See A.2.c.(4), below, for the information the CSHO shall attempt to obtain.)

(a)   If the employer refuses to allow an inspection of the establishment to proceed, the CSHO shall leave the premises and immediately report the refusal to the Assistant Area Director. The Area Director shall notify the Regional Solicitor.

(b)   If the employer raises no objection to inspection of certain portions of the workplace but objects to inspection of other portions, this shall be documented. Normally, the CSHO shall continue the inspection, confining it only to those certain portions to which the employer has raised no objections.

(c)   In either case the CSHO shall advise the employer that the refusal will be reported to the Assistant Area Director and that the agency may take further action, which may include obtaining legal process.

(d)   On multiemployer worksites, valid consent can be granted by the owner, or another co-occupier of the space, for site entry.

(2)   <u>Employer Interference</u>. Where entry has been allowed but the employer interferes with or limits any important aspect of the inspection, the CSHO shall determine whether or not to consider this action as a refusal. Examples of interference are refusals to permit the walkaround, the examination of records essential to the inspection, the taking of essential photographs and/or videotapes, the inspection of a particular

part of the premises, indispensable employee interviews, or the refusal to allow attachment of sampling devices.

A. 2. c.   (3)   <u>Administrative Subpoena</u>. Whenever there is a reasonable need for records, documents, testimony and/or other supporting evidence necessary for completing an inspection scheduled in accordance with any current and approved inspection scheduling system or an investigation of any matter properly falling within the statutory authority of the agency, the Regional Administrator, or authorized Area Director, may issue an administrative subpoena. (See OSHA Instruction ADM 4.4.)

  (4)   <u>Obtaining Compulsory Process</u>. If it is determined, upon refusal of entry or refusal to produce evidence required by subpoena, that a warrant will be sought, the Area Director shall proceed according to guidelines and procedures established in the Region for warrant applications.

    (a)   With the approval of the Regional Solicitor, the Area Director may initiate the compulsory process.

    (b)   The warrant sought when employer consent has been withheld shall normally be limited to the specific working conditions or practices forming the basis of the unprogrammed inspection. A broad scope warrant, however, may be sought when the information available indicates conditions which are pervasive in nature or if the establishment is on the current list of targeted establishments.

    (c)   If the warrant is to be obtained by the Regional Solicitor, the Area Director shall transmit in writing to the Regional Solicitor, within 48 hours after the determination is made that a warrant is necessary, all information necessary to obtain a warrant, as determined through contact with the Solicitor, which may include the following:

      <u>1</u>   Area/District Office, telephone number, and name of Assistant Area Director involved.

      <u>2</u>   Name of CSHO attempting inspection and inspection number, if assigned. Identify whether inspection to be conducted included safety items, health items or both.

      <u>3</u>   Legal name of establishment and address including City, State and County. Include site location if different from mailing address.

A. 2. c.   (4)   (c)   <u>4</u>   Estimated number of employees at inspection site.

                 <u>5</u>   SIC Code and high hazard ranking for that specific industry within the State, as obtained from statistics provided by the National Office.

                 <u>6</u>   Summary of all facts leading to the refusal of entry or limitation of inspection, including the following:

                       <u>a</u>   Date and time of entry.

                       <u>b</u>   Date and time of denial.

                       <u>c</u>   Stage of denial (entry, opening conference, walkaround, etc.).

                 <u>7</u>   Narrative of all actions taken by the CSHO leading up to during and after refusal, including the following information:

                       <u>a</u>   Full name and title of the person to whom CSHO presented credentials.

                       <u>b</u>   Full name and title of person(s) who refused entry.

                       <u>c</u>   Reasons stated for the denial by person(s) refusing entry.

                       <u>d</u>   Response, if any, by CSHO to <u>c</u>, above.

                       <u>e</u>   Name and address of witnesses to denial of entry.

                 <u>8</u>   All previous inspection information, including copies of the previous citations.

                 <u>9</u>   Previous requests for warrants. Attach details, if applicable.

                 <u>10</u>   As much of the current inspection report as has been completed.

                 <u>11</u>   If a construction site involving work under contract from any agency of the Federal Government, the name of the agency, the date of the contract, and the type of work involved.

                 <u>12</u>   Other pertinent information such as description of the workplace; the work process; machinery, tools and materials used;

known hazards and injuries associated with the specific manufacturing process or industry.

A. 2. c. (4) (c) <u>13</u> Investigative techniques which will be required during the proposed inspection; e.g., personal sampling, photographs, audio/videotapes, examination of records, access to medical records, etc.

<u>14</u> The specific reasons for the selection of this establishment for the inspection including proposed scope of the inspection and rationale:

<u>a</u> <u>Imminent Danger</u>.

o Description of alleged imminent danger situation.

o Date received and source of information.

o Original allegation and copy of typed report, including basis for reasonable expectation of death or serious physical harm and immediacy of danger.

o Whether all current imminent danger processing procedures have been strictly followed.

<u>b</u> <u>Fatality/Catastrophe</u>.

o The OSHA-36F filled out in as much detail as possible.

<u>c</u> <u>Complaint or Referral</u>.

o Original complaint or referral and copy of typed complaint or referral.

o Reasonable grounds for believing that a violation that threatens physical harm or imminent danger exists, including standards that could be violated if the complaint or referral is true and accurate.

o Whether all current complaint or referral processing procedures have been strictly followed.

o Additional information gathered pertaining to complaint or referral evaluation.

A. 2. c. (4) (c) <u>14</u> <u>d</u>   <u>Programmed</u>.

o   Targeted safety--general industry, maritime, construction.

o   Targeted health.

o   Special emphasis program--Special Programs, Local
Emphasis Program, Migrant Housing Inspection, etc.

<u>e</u>   <u>Followup</u>.

o   Date of initial inspection.

o   Details and reasons followup was to be conducted.

o   Copies of previous citations on the basis of which
the followup was initiated.

o   Copies of settlement stipulations and final orders, if
appropriate.

o   Previous history of failure to correct, if any.

<u>f</u>   <u>Monitoring</u>.

o   Date of original inspection.

o   Details and reasons monitoring inspection was to be
conducted.

o   Copies of previous citations and/or settlement
agreements on the basis of which the monitoring
inspection was initiated.

o   PMA request, if applicable.

(5)   <u>Compulsory Process</u>. When a court order or warrant is obtained requiring an employer to allow an inspection, the CSHO is authorized to
conduct the inspection in accordance with the provisions of the court
order or warrant. All questions from the employer concerning reason-
ableness of any aspect of an inspection conducted pursuant to compulsory process shall be referred to the Area Director.

A. 2. c. (6) Action to be Taken Upon Receipt of Compulsory Process. The inspection will normally begin within 24 hours of receipt of a warrant or of the date authorized by the warrant for the initiation of the inspection.

(a) The CSHO shall serve a copy of the warrant on the employer and make a separate notation as to the time, place, name and job title of the individual served.

(b) The warrant may have a space for a return of service entry by the CSHO in which the exact dates of the inspection made pursuant to the warrant are to be entered. Upon completion of the inspection, the CSHO will complete the return of service on the original warrant, sign and forward it to the Assistant Area Director for appropriate action.

(c) Even where the walkaround is limited by a warrant or an employer's consent to specific conditions or practices, a subpoena for production of records shall be normally served, if necessary, in accordance with A.2.c.(3), above. The records specified in the subpoena shall include (as appropriate) injury and illness records, exposure records, the written hazards communication program, the written lockout-tagout program, and records relevant to the employer's safety and health management program, such as safety and health manuals or minutes from safety meetings.

(d) The Regional Administrator, or Area Director authorized to do so, may issue, for each inspection, an administrative subpoena which seeks production of the above specified categories of documents. The subpoena may call for immediate production of the records with the exception of the documents relevant to the safety and health management program, for which a period of 5 working days normally shall be allowed.

(e) If circumstances make it appropriate, a second warrant may be sought based on the review of records or on "plain view" observations of other potential violations during a limited scope walkaround.

(7) Federal Marshal Assistance. A U.S. Marshal may accompany the CSHO when the compulsory process is presented.

A. 2. c. (8) <u>Refused Entry or Interference with a Compulsory Process</u>.

(a) When an apparent refusal to permit entry or inspection is encountered upon presenting the warrant, the CSHO shall specifically inquire whether the employer is refusing to comply with the warrant.

(b) If the employer refuses to comply or if consent is not clearly given, the CSHO shall not attempt to conduct the inspection but shall leave the premises and contact the Assistant Area Director concerning further action. The CSHO shall make notations (including all witnesses to the refusal or interference) and fully report all relevant facts. Under these circumstances the Area Director shall contact the Regional Solicitor and they shall jointly decide what further action shall be taken.

d. <u>Forcible Interference with Conduct of Inspection or Other Official Duties</u>. Whenever an OSHA official or employee encounters forcible resistance, opposition, interference, etc., or is assaulted or threatened with assault while engaged in the performance of official duties, all investigative activity shall cease.

(1) The Assistant Area Director shall be advised by the most expeditious means.

(2) Upon receiving a report of such forcible interference, the Area Director or designee shall immediately notify the Regional Administrator.

e. <u>Release for Entry</u>.

(1) The CSHO shall not sign any form or release or agree to any waiver. This includes any employer forms concerned with trade secret information.

(2) The CSHO may obtain a pass or sign a visitor's register, or any other book or form used by the establishment to control the entry and movement of persons upon its premises. Such signature shall not constitute any form of a release or waiver of prosecution of liability under the Act.

f. <u>Bankrupt or Out of Business</u>. If the establishment scheduled for inspection is found to have ceased business and there is no known successor, the CSHO shall report the facts to the Assistant Area Director. If an employer, although adjudicated bankrupt, is continuing to operate on the date of the scheduled inspection, the inspection shall proceed. An employer must comply with the Act until the day the business actually ceases to operate.

A. 2. g. <u>Strike or Labor Dispute</u>. Plants or establishments may be inspected regardless of the existence of labor disputes involving work stoppages, strikes or picketing. If the CSHO identifies an unanticipated labor dispute at a proposed inspection site, the Assistant Area Director shall be consulted before any contact is made.

(1) <u>Programmed Inspections</u>. Programmed inspections may be deferred during a strike or labor dispute, either between a recognized union and the employer or between two unions competing for bargaining rights in the establishment.

(2) <u>Unprogrammed Inspections</u>. Unprogrammed inspections (complaints, fatalities, etc.) will be performed during strikes or labor disputes. However, the seriousness and reliability of any complaint shall be thoroughly investigated by the supervisor prior to scheduling an inspection to ensure as far as possible that the complaint reflects a good faith belief that a true hazard exists. If there is a picket line at the establishment, the CSHO shall inform the appropriate union official of the reason for the inspection prior to initiating the inspection.

h. <u>Employee Participation</u>. The CSHO shall advise the employer that Section 8(e) of the Act and 29 CFR §1903.8 require that an employee representative be given an opportunity to participate in the inspection.

(1) CSHOs shall determine as soon as possible after arrival whether the employees at the worksite to be inspected are represented and, if so, shall ensure that employee representatives are afforded the opportunity to participate in all phases of the workplace inspection.

(2) If an employer resists or interferes with participation by employee representatives in an inspection and this cannot be resolved by the CSHO, the continued resistance shall be construed as a refusal to permit the inspection and the Assistant Area Director shall be contacted in accordance with A.2.c. of this chapter.

NOTE: For the purpose of this chapter, the term "employee representative" refers to (1) a representative of the certified or recognized bargaining agent, or, if none, (2) an employee member of a safety and health committee who has been chosen by the employees (employee committee members or employees at large) as their

OSHA representative, or (3) an individual employee who has been selected as the walkaround representative by the employees of the establishment.

A. 3. Opening Conference. The CSHO shall attempt to inform all effected employers of the purpose of the inspection, provide a copy of the complaint if applicable, and shall include employees unless employer objects. The opening conference shall be kept as brief as possible and may be expedited through use of an opening conference handout. Conditions of the worksite shall be noted upon arrival as well as any changes which may occur during the opening conference.

NOTE: The CSHO shall determine if the employer is covered by any of the exemptions or limitations noted in the current Appropriations Act (see OSHA Instruction CPL 2.51H, or the most current version), or deletions in OSHA Instruction CPL 2.45B Chapter II, F.2.b.(1)(b)5 b or superseding directive.

a. Attendance at Opening Conference. OSHA encourages employers and employees to meet together in the spirit of open communication. The CSHO shall conduct a joint opening conference with employer and employee representatives unless either party objects. If there is objection to a joint conference, the CSHO shall conduct separate conferences with employer and employee representatives.

b. Scope. The CSHO shall outline in general terms the scope of the inspection, including private employee interviews, physical inspection of the workplace and records, possible referrals, discrimination complaints, and the closing conference(s).

c. Forms Completion. The CSHO shall obtain available information for the OSHA-1 and other appropriate forms.

d. Employees of Other Employers. During the opening conference, the CSHO shall determine whether the employees of any other employers are working at the establishment. If these employers may be affected by the inspection, the scope may be expanded to include others or a referral made at the discretion of the CSHO. At multiemployer sites, copies of complaint(s), if applicable, shall be provided to all employers affected by the alleged hazard(s), and to the general contractor.

e. Voluntary Compliance Programs. Employers who participate in selected voluntary compliance programs may be exempted from programmed inspections. The CSHO shall determine whether the employer falls under such an exemption during the opening conference.

A. 3. e. (1) <u>Section 7(c)(1) and Contract Consultations</u>. In accordance with 29 CFR §1908.7 and Chapter IX of the Consultation Policies and Procedures Manual (CPPM), the CSHO shall ascertain at the opening conference whether an OSHA-funded consultation is in progress or whether the facility is pursuing or has received an inspection exemption through consultation under current procedures.

      (a) An on site consultation visit in progress has priority over programmed inspections except as indicated in 29 CFR §1908.7(b)(2)(iv), which allows for critical inspections as determined by the Assistant Secretary.

      (b) If a consultation visit is in progress, the inspection may be rescheduled.

      (c) If a followup inspection (including monitoring) or an imminent danger, fatality/catastrophe, complaint or referral investigation is to be conducted, the inspection shall not be deferred, but its scope shall be limited to those areas required to complete the purpose of the investigation. The consultant must interrupt the onsite visit until the compliance inspection shall have been completed (Ref. 29 CFR §1908.7).

  (2) <u>Voluntary Protection Programs (VPP)</u>. In the event a CSHO enters a facility that has been approved for participation in a VPP and is currently under an inspection exemption, the approval letter shall be copied and the inspection either be terminated (if it is a programmed inspection) or limited to the specified items in the complaint or referral (if it is unprogrammed).

  f. <u>Walkaround Representatives</u>. Those representatives designated to accompany the CSHO during the walkaround are considered walkaround representatives, and will generally include employer designated and employee designated representatives. At establishments where more than one employer is present or in situations where groups of employees have different representatives, it is acceptable to have a different employer/employee representative for different phases of the inspection. More than one employer and/or employee representative may accompany the CSHO throughout or during any phase of an inspection if the CSHO determines that such additional representatives will aid, and not interfere with, the inspection (29 CFR §1903.8(a)).

  (1) <u>Employees Represented by a Certified or Recognized Bargaining Agent</u>. During the opening conference, the highest ranking union official or union employee representative on-site shall designate who will participate in the walkaround. OSHA regulation 29 CFR §1903.8(b)

gives the CSHO the authority to resolve all disputes as to who is the representative authorized by the employer and employees. Title 29 CFR §1903.8(c) states that the representative authorized by the employees shall be an employee of the employer. The CSHO can decide to include others.

A. 3. f. (2) Safety Committee. The employee members of an established plant safety committee or the employees at large may have designated an employee representative for OSHA inspection purposes or agreed to accept as their representative the employee designated by the committee to accompany the CSHO during an OSHA inspection.

(3) No Certified or Recognized Bargaining Agent. Where employees are not represented by an authorized representative, where there is no established safety committee, or where employees have not chosen or agreed to an employee representative for OSHA inspection purposes whether or not there is a safety committee, the CSHO shall determine if any other employees would suitably represent the interests of employees on the walkaround. If selection of such an employee is impractical, the CSHO shall consult with a reasonable number of employees during the walkaround.

g. Preemption by Another Agency. Section 4(b)(1) of the OSH Act states that the OSH Act does not apply to working conditions over which other Federal agencies exercise statutory responsibility. The determination of preemption by another Federal agency is, in many cases, a highly complex matter. Any such situations shall be brought to the attention of the Area Director as soon as they arise, and dealt with on a case by case basis.

h. Disruptive Conduct. The CSHO may deny the right of accompaniment to any person whose conduct interferes with a full and orderly inspection (29 CFR §1903.8(d)). If disruption or interference occurs, the CSHO shall use professional judgment as to whether to suspend the walkaround or take other action. The Assistant Area Director shall be consulted if the walkaround is suspended. The employee representative shall be advised that during the inspection matters unrelated to the inspection shall not be discussed with employees.

i. Trade Secrets. The CSHO shall ascertain from the employer if the employee representative is authorized to enter any trade secret area(s). If not, the CSHO shall consult with a reasonable number of employees who work in the area (29 CFR §1903.9(d)).

A. 3. j.   Classified Areas. In areas containing information classified by an agency of the U.S. Government in the interest of national security, only persons authorized to have access to such information may accompany a CSHO (29 CFR §1903.8(d)).

     k.   Examination of Record Programs and Posting Requirements.

          (1)   Records. As appropriate, the CSHO shall review the injury and illness records to the extent necessary to determine compliance and identify trends. Other OSHA programs and records will be reviewed at the CSHO's professional discretion as necessary.

          (2)   Lost Workday Injury (LWDI) Rate. The LWDI may in the CSHO's discretion be used in determining trends in injuries and illnesses. The LWDI rate is calculated according to the following formula:

          If the number of employees hours worked is available from the employer, use:

          $$\text{LWDI Rate} = \frac{\text{\# LWDI's} \times 200{,}000}{\text{\# employee hours worked}}$$

          Where:

               # LWDI's = sum of LWDI's in the reference years.

               # employee hours worked = sum of employee hours in the reference years.

               200,000 = base for 100 full-time workers, working 40 hours per week, 50 weeks per year.

          EXAMPLE: An establishment scheduled for inspection in October 1993 employed an average of 54 workers in 1992, 50 workers in 1991, and 50 workers in 1990. Therefore, injury and employment data for the two preceding calendar years will be used.

               # LWDI's in 1991 = 5

               # LWDI's in 1992 = 3

               # Employee hours worked in 1991 = 100,000

               # Employee hours worked in 1992 = 108,000

II-13

$$\text{LWDI Rate} \quad = \quad \frac{(5 + 3) \times 200{,}000}{100{,}000 + 108{,}000}$$

$$= \quad \frac{1{,}600{,}000}{208{,}000}$$

$$= \quad 7.69 \text{ (rounded to 7.7)}$$

(3) <u>Posting</u>. The CSHO shall determine if posting requirements are met in accordance with 29 CFR Parts 1903 and 1904.

A. 4. <u>Walkaround Inspection</u>. The main purpose of the walkaround inspection is to identify potential safety and/or health hazards in the workplace. The CSHO shall conduct the inspection in such a manner as to eliminate unnecessary personal exposure to hazards and to minimize unavoidable personal exposure to the extent possible.

a. <u>Evaluation</u>. The employer's safety and health program shall be evaluated to determine the employer's good faith. See Chapter IV, C.2.i.(5)(b).

b. <u>Record All Facts Pertinent to an Apparent Violation</u>. Apparent violations shall be brought to the attention of employer and employee representatives at the time they are documented. CSHOs shall record at a minimum the identity of the exposed employee, the hazard to which the employee was exposed, the employee's proximity to the hazard, the employer's knowledge of the condition, and the manner in which important measures were obtained.

NOTE: If employee exposure (either to safety or health hazards) is not observed, the CSHO shall document facts on which the determination is made that an employee has been or could be exposed.

c. <u>Collecting Samples</u>.

(1) The CSHO shall determine as soon as possible after the start of the inspection whether sampling, such as but not limited to air sampling and surface sampling, is required by utilizing the information collected during the walkaround and from the pre-inspection review.

(2) If either the employer or the employee representative requests sampling results, summaries of the results shall be provided to the requesting representative as soon as practicable.

(3) The CSHO may reference the sampling strategy located in OSHA Instruction CPL 2-2.20B for additional information on sampling techniques.

A. 4. d.   Taking Photographs and/or Videotapes. Photographs and/or videotapes shall be taken whenever the CSHO judges there is a need. Photographs that support violations shall be properly labeled, and may be attached to the appropriate OSHA-1B. The CSHO shall ensure that any photographs relating to confidential or trade secret information are identified as such. All film and photographs shall be retained in the case file. Videotapes shall be properly labeled and stored. Refer to OSHA Instruction CPL 2.98 for further information on videotaping.

e.   Interviews. A free and open exchange of information between the CSHO and employees is essential to an effective inspection. Interviews provide an opportunity for employees or other individuals to point out hazardous conditions and, in general, to provide assistance as to what violations of the Act may exist and what abatement action should be taken. Employee interviews are also an effective means to determine if advance notice of inspection, when given under the guidelines in Chapter I, E.3., has adversely affected the inspection conditions.

(1)   Purpose. Section 8(a)(2) of the Act authorizes the CSHO to question any employee privately during regular working hours in the course of an OSHA inspection. The purpose of such interviews is to obtain whatever information the CSHO deems necessary or useful in carrying out the inspection effectively. Such interviews shall be kept as brief as possible. Individual interviews are authorized even when there is an employee representative present.

(2)   Employee Right of Complaint. The CSHO may consult with any employee who desires to discuss a possible violation. Upon receipt of such information, the CSHO shall investigate the alleged hazard, where possible, and record the findings. If a written complaint is received, the written response procedures in Chapter I shall be followed.

(3)   Time and Location. Interviews shall be conducted in a reasonable manner and normally will be conducted during the walkaround; however, they may be conducted at any time during an inspection. If necessary, interviews may be conducted at locations other than the workplace.

(4)   Privacy. Employers shall be informed that the interview is to be in private. Whenever an employee expresses a preference that an employee representative be present for the interview, the CSHO shall

make a reasonable effort to honor that request. Any employer objection to private interviews with employees may be construed as a refusal of entry and handled in accordance with the procedures in A.2.c. of this chapter.

A. 4. e. (5) <u>Interview Statements</u>. Interview statements of employees or other individuals shall be obtained whenever the CSHO determines that such statements would be useful in documenting adequately an apparent violation.

    (a)   Interviews shall normally be reduced to writing, and the individual shall be encouraged to sign and date the statement. The CSHO shall assure the individual that the statement will be held confidential to the extent allowed by law, but they may be used in court/hearings. See OSHA Instruction CPL 2.98 for guidance on videotaping.

    (b)   Interview statements shall normally be written in the first person and in the language of the individual.

        <u>1</u>   Any changes or corrections shall be initialed by the individual; otherwise, the statement shall not be changed, added to or altered in any way.

        <u>2</u>   The statements shall end with wording such as: "I have read the above, and it is true to the best of my knowledge." The statement shall also include the following: "I request that my statement be held confidential to the extent allowed by law." The individual,
however, may waive confidentiality. The individual shall sign and date the statement and the CSHO shall then sign it as a witness.

        <u>3</u>   If the individual refuses to sign the statement, the CSHO shall note such refusal on the statement. The statement shall, nevertheless, be read to the individual and an attempt made to obtain agreement. A note that this was done shall be entered into the case file.

    (c)   A transcription of a recorded statement shall be made if necessary.

    f.   <u>Employer Abatement Assistance</u>.

    (1)   <u>Policy</u>. CSHOs shall offer appropriate abatement assistance during the walkaround as to how workplace hazards might be eliminated. The

information shall provide guidance to the employer in developing acceptable abatement methods or in seeking appropriate professional assistance. CSHO's shall not imply OSHA endorsement of any product through use of specific product names when recommending abatement measures. The issuance of citations shall not be delayed.

A. 4. f.   (2) <u>Disclaimers</u>. The employer shall be informed that:

     (a) The employer is not limited to the abatement methods suggested by OSHA;

     (b) The methods explained are general and may not be effective in all cases; and

     (c) The employer is responsible for selecting and carrying out an effective abatement method.

  g. <u>Special Circumstances</u>.

    (1) <u>Trade Secrets</u>. Trade secrets are matters that are not of public or general knowledge. A trade secret is any confidential formula, pattern, process, equipment, list, blueprint, device or compilation of information used in the employer's business which gives an advantage over competitors who do not know or use it.

     (a) <u>Policy</u>. It is essential to the effective enforcement of the Act that the CSHO and all OSHA personnel preserve the confidentiality of all information and investigations which might reveal a trade secret.

     (b) <u>Restrictions and Controls</u>. When the employer identifies an operation or condition as a trade secret, it shall be treated as such. Information obtained in such areas, including all negatives, photographs, videotapes, and OSHA documentation forms, shall be labeled:

"ADMINISTRATIVELY CONTROLLED INFORMATION"
"RESTRICTED TRADE INFORMATION"

      <u>1</u> Under Section 15 of the Act, all information reported to or obtained by a CSHO in connection with any inspection or other activity which contains or which might reveal a trade secret shall be kept confidential. Such information shall not be disclosed except to other OSHA officials concerned with the enforcement of the Act or, when relevant, in any proceeding under the Act.

A. 4. g.   (1) (b)   <u>2</u>   Title 18 of the United States Code, Section 1905, provides criminal penalties for Federal employees who disclose such information. These penalties include fines of up to $1,000 or imprisonment of up to one year, or both, and removal from office or employment.

                             <u>3</u>   Trade secret materials shall not be labeled as "Top Secret," "Secret," or "Confidential," nor shall these security classification designations be used in conjunction with other words unless the trade secrets are also classified by an agency of the U.S. Government in the interest of national security.

            (c)   <u>Photographs and Videotapes</u>. If the employer objects to the taking of photographs and/or videotapes because trade secrets would or may be disclosed, the CSHO should advise the employer of the protection against such disclosure afforded by Section 15 of the Act and 29 CFR §1903.9. If the employer still objects, the CSHO shall contact the Assistant Area Director.

      (2)   <u>Violations of Other Laws</u>. If a CSHO observes apparent violations of laws enforced by other government agencies, such cases shall be referred to the appropriate agency. Referrals shall be made using appropriate Regional procedures (see A.3.g. of this chapter).

  5.   <u>Closing Conference</u>.

      a.   At the conclusion of an inspection, the CSHO shall conduct a closing conference with the employer and the employee representatives, jointly or separately, as circumstances dictate. The closing conference may be conducted on site or by telephone as deemed appropriate by the CSHO.

          NOTE: When conducting separate closing conferences for employers and labor representatives (where the employer has declined to have a joint closing conference with employee representatives), the CSHO shall normally hold the conference with employee representatives first, unless the employee representative requests otherwise. This procedure will ensure that worker input, if any, is received--and that any needed changes are made--before employers are informed of violations and proposed citations.

      b.   The CSHO shall describe the apparent violations found during the inspection and other pertinent issues as deemed necessary by the CSHO. Both the employer and the employee representatives shall be advised of their rights to participate in any subsequent conferences, meetings or discussions, and their contest rights. Any unusual circumstances noted during the closing conference shall be documented in the case file.

A. 5. b. (1) Since the CSHO may not have all pertinent information at the time of the first closing conference, a second closing conference may be held by telephone or in person to inform the employer and the employee representatives whether the establishment is in compliance.

(2) The CSHO shall advise the employee representatives that:

(a) Under 29 CFR § 2200.20 of the Occupational Safety and Health Review Commission regulations, if the employer contests, the employees have a right to elect "party status" before the Review Commission.

(b) They must be notified by the employer if a notice of contest or a petition for modification of abatement date is filed.

(c) They have Section 11(c) rights.

(d) They have a right to contest the abatement date. Such contest must be in writing and must be filed within 15 working days after receipt of the citation.

B. Special Inspection Procedures.

1. Followup and Monitoring Inspections.

a. Inspection Procedures. The primary purpose of a followup inspection is to determine if the previously cited violations have been corrected. Monitoring inspections are conducted to ensure that hazards are being corrected and employees are being protected, whenever a long period of time is needed for an establishment to come into compliance, or to verify compliance with the terms of granted variances. Issuance of willful, repeated and high gravity serious violations, failure to abate notifications, and/or citations related to imminent danger situations are examples of prime candidates for followup or monitoring inspections. Followup or monitoring inspections would not normally be conducted when evidence of abatement is provided by the employer or employee representatives. Normally, there shall be no additional inspection activity unless, in the judgment of the CSHO, there have been significant changes in the workplace which warrant further inspection activity.

B. 1. b. <u>Failure to Abate</u>.

(1) A failure to abate exists when the employer has not corrected a violation for which a citation has been issued and abatement date has passed or which is covered under a settlement agreement, or has not complied with interim measures involved in a long-term abatement within the time given.

(2) If the cited items have not been abated, a Notice of Failure to Abate Alleged Violation shall normally be issued. If a subsequent inspection indicates the condition has still not been abated, the Regional Solicitor shall be consulted for further guidance.

> **NOTE:** If the employer has exhibited good faith, a late PMA may be considered in accordance with Chapter IV, D.2. where there are extenuating circumstances.

(3) If it is determined that the originally cited violation was abated but then recurred, a citation for repeated violation may be appropriate.

c. <u>Reports</u>.

(1) A copy of the previous OSHA-1B, OSHA-1BIH, or citation can be used, and "corrected" written on it, with a brief explanation of the correction if deemed necessary by the CSHO, for those items found to be abated. This information may alternately be included in the narrative or in video/audio documentation.

(2) In the event that any item has not been abated, complete documentation shall be included on an OSHA-1B.

d. <u>Followup Files</u>. The followup inspection reports shall be included with the original (parent) case file.

2. <u>Fatality/Catastrophe Investigations</u>. For guidance on conducting fatality and catastrophe inspections, refer to OSHA Instruction CPL 2.77 and CPL 2.94.

a. <u>Definitions</u>. The following definitions apply for purposes of this section:

(1) <u>Fatality</u>. An employee death resulting from a work-related incident or exposure; in general, from an accident or illness caused by or related to a workplace hazard.

B. 2. a. (2) <u>Catastrophe</u>. The hospitalization of three or more employees resulting from a work-related incident; in general, from an accident or illness caused by a workplace hazard.

(3) <u>Hospitalization</u>. To be admitted as an <u>inpatient</u> to a hospital or equivalent medical facility for examination or treatment.

(4) <u>Reporting</u>. Area Directors shall report all job-related fatalities and catastrophes which may result in high media attention or have national implications and that appear to be within OSHA's jurisdiction as soon as they become aware of them to the Regional Administrator. See CPL 2.97.

b. <u>Selection of CSHO</u>. A CSHO, preferably with expertise in the particular industry or operation involved in the accident or illness, shall be selected by the Area Director and sent to the establishment as soon as possible. If a potential criminal violation appears possible during the inspection, staff who have received criminal investigation training at the Federal Law Enforcement Training Center shall be assigned, if available.

c. <u>Families of Victims</u>.

(1) Family members of employees involved in fatal occupational accidents or illnesses shall be contacted at an early point in the investigation, given an opportunity to discuss the circumstances of the accident or illness, and provided timely and accurate information at all stages of the investigations as directed in (2), below.

(2) All of the following require special tact and good judgment on the part of the CSHO. In some situations, these procedures should not be followed to the letter; e.g., in some small businesses, the employer, owner, or supervisor may be a relative of the victim. In such circumstances, such steps as issuance of the form letter may not be appropriate without some editing.

(a) As soon as practicable after initiating the investigation, the CSHO shall attempt to compile a list of all of the accident victims and their current addresses, along with the names of individual(s) listed in the employer's records as next-of-kin (family member(s)) or person(s) to contact in the event of an emergency.

(b) The standard information letter should be sent to the family member(s) or the person(s) listed as the emergency contact person(s) indicated on the victims' employment records within 5 working days of the time their identities have been established.

(c)   The compliance officer, when taking a statement from families of victims, shall explain that the interview will be kept confidential to the extent allowed by law and that the interview will be handled following the same procedures as employee interviews. The greatest sensitivity and professionalism is required for such an interview. The information received must be carefully evaluated and corroborated during the investigation.

(d)   Followup contact shall be maintained with a key family member or other contact person, when requested, so that the survivors can be kept up-to-date on the status of the investigation. The victim's family members shall be provided a copy of all citations issued as a result of the accident investigation within 5 working days of issuance.

B. 2. d.   Criminal. Section 17(e) of the Act provides criminal penalties for an employer who is convicted of having willfully violated an OSHA standard, rule or order when that violation caused the death of an employee. In an investigation of this type, therefore, the nature of the evidence available is of paramount importance. There shall be early and close liaison between the OSHA investigator, the Area Director, the Regional Administrator and the Regional Solicitor in developing any finding which might involve a violation of Section 17(e) of the Act. An OSHA investigator with criminal investigation training shall be assigned at an early stage to assist in developing the case.

e.   Rescue Operations. OSHA has no authority to direct rescue operations--this is the responsibility of the employer and/or of local political subdivisions or State agencies. OSHA does have the authority to monitor and inspect the working conditions of covered employees engaged in rescue operations to make certain that all necessary procedures are being taken to protect the lives of the rescuers. See also memorandum on Policy Regarding Voluntary Rescue Activities, dated March 31, 1994, to the Regional Administrators from H. Berrien Zettler, Deputy Director, Directorate of Compliance Programs

f.   Public Information Policy. The OSHA public information policy regarding response to fatalities and catastrophes is to explain Federal presence to the news media. It is not to provide a continuing flow of facts nor to issue periodic updates on the progress of the investigation. The Area Director or his/her designee shall normally handle responses to media inquiries.

3.   Imminent Danger Investigations.

a.   Definition. Section 13(a) of the Act defines imminent danger as ". . . any conditions or practices in any place of employment which are such that a danger exists which could reasonably be expected to cause death or serious physical harm immediately or before the imminence of such danger can be eliminated through the enforcement procedures otherwise provided by this Act."

B. 3. b.   Requirements. The following conditions must be met before a hazard becomes an imminent danger:

(1)   It must be reasonably likely that a serious accident will occur immediately (see B.3.c.(2)(b), below) or, if not immediately, then before abatement would otherwise be required (see B.3.c.2.(c), below). If an employer contests a citation, abatement will not be required until there is a final order of the Review Commission affirming the citation.

(2)   The harm threatened must be death or serious physical harm. For a health hazard, exposure to the toxic substance or other health hazard must cause harm to such a degree as to shorten life or cause substantial reduction in physical or mental efficiency even though the resulting harm may not manifest itself immediately.

c.   Inspection.

(1)   Scope. CSHO may consider expanding the scope of inspection based on the information available during the inspection process.

(2)   Elimination of the Imminent Danger. As soon as reasonably practicable after it is concluded that conditions or practices exist which constitute an imminent danger, the employer shall be so advised and requested to notify its employees of the danger and remove them from exposure to the imminent danger. The employer should be encouraged to do whatever is possible to eliminate the danger promptly on a voluntary basis.

(a)   Voluntary Elimination of the Imminent Danger. The employer may voluntarily and permanently eliminate the imminent danger as soon as it is pointed out. In such cases, no imminent danger proceeding need be instituted; and, no Notice of Alleged Imminent Danger completed. An appropriate citation and notification of penalty shall be issued.

(b)   Action Where the Danger is Immediate and Voluntary Elimination Is Not Accomplished. If the employer either cannot or does not voluntarily eliminate the hazard or remove employees from the exposure and the danger is immediate, the following procedures shall be observed:

B. 3. c. (2) (b) <u>1</u>   The CSHO shall post the OSHA-8 and call the Area Director, who will decide whether to contact the Regional Solicitor to obtain a Temporary Restraining Order (TRO). The Regional Administrator shall be notified of the TRO proceedings.

> **NOTE:**   The CSHO has no authority to order the closing of the operation or to direct employees to leave the area of the imminent danger or the workplace.

        <u>2</u>   The CSHO shall notify employees and employee representatives of the posting of the OSHA-8 and shall advise them of their Section 11(c) rights.

        <u>3</u>   The employer shall be advised that Section 13 of the Act gives United States District Courts jurisdiction to restrain any condition or practice which is an imminent danger to employees.

        <u>4</u>   The Area Director and the Regional Solicitor shall assess the situation and make arrangements for the expedited initiation of court action, if warranted, or instruct the CSHO to remove the OSHA-8.

        <u>5</u>   The CSHO's first priority in scheduling activities is to prepare for litigation related to TRO's in imminent danger matters.

    (c)   <u>Action Where the Danger is that the Harm will Occur Before Abatement is Required</u>. If the danger is that the harm will occur before abatement is required, i.e. before a final order of the Commission can be obtained in a contested case, the CSHO shall contact the Area Director and Regional Solicitor.

        <u>1</u>   In many cases, the CSHO or the AD may not decide there is such an imminent danger at the time of the physical inspection of the plant. Further evaluation of the file or additional evidence may warrant consultation with the Regional Solicitor.

        <u>2</u>   In appropriate cases, the imminent danger notice may be posted at the time citations are delivered or even after the notice of contest is filed.

B. 4. <u>Construction Inspections</u>.

    a. <u>Standards Applicability</u>. The standards published as 29 CFR Part 1926 have been adopted as occupational safety and health standards under Section 6(a) of the Act and 29 CFR §1910.12. They shall apply to every employment and place of employment of every employee engaged in construction work, including non-contract construction work.

    b. <u>Definition</u>. The term "construction work" means work for construction, alteration, and/or repair, including painting and decorating. These terms are discussed in 29 CFR §1926.13. If any question arises as to whether an activity is deemed to be construction for purposes of the Act, the Director of Compliance Programs shall be consulted.

    c. <u>Employer Worksite</u>.

        (1) <u>General</u>. Inspections of employers in the construction industry are not easily separable into distinct worksites. The worksite is generally the site where the construction is being performed (e.g., the building site, the dam site). Where the construction site extends over a large geographical area (e.g., road building), the entire job will be considered a single worksite. In cases when such large geographical areas overlap between Area Offices, generally only operations of the employer within the jurisdiction of any Area Office will be considered as the worksite of the employer.

        (2) <u>Beyond Single Area Office</u>. When a construction worksite extends beyond a single Area Office and the CSHO believes that the inspection should be extended, the affected Area Directors shall consult with each other and take appropriate action.

B. 4. d. <u>Entry of the Workplace</u>.

        (1) <u>Other Agency</u>. The CSHO shall ascertain whether there is a representative of a Federal contracting agency at the worksite. If so, the CSHO shall contact the representative, advise him/her of the inspection and request that he/she attend the opening conference. (For Federal Agencies, see Chapter XIII and following Appendix A, of OSHA Instruction CPL 2.45B or a superseding directive).

        (2) <u>Complaints</u>. If the inspection is being conducted as a result of a complaint, a copy of the complaint is to be furnished to the general contractor and any affected sub-contractors.

e. <u>Closing Conference</u>. Upon completion of the inspection, the CSHO shall confer with the general contractors and all appropriate subcontractors or their representatives, together or separately, and advise each one of all the apparent violations disclosed by the inspection to which each one's employees were exposed, or violations which the employer created or controlled. Employee representatives participating in the inspection shall also be afforded the right to participate in the closing conference(s).

B. 5. <u>Federal Agency Inspections</u>. Policies and procedures for Federal agencies are to be the same as those followed in the private sector, except as specified in Chapter XIII, and the following Appendix A, of OSHA Instruction CPL 2.45B or a superseding directive.

CHAPTER III

INSPECTION DOCUMENTATION

A. <u>Four Stage Case File Documentation</u>.

   1. <u>General</u>.

      a. <u>Guidelines</u>. These guidelines are developed to assist the CSHO in determining the minimum level of written documentation appropriate for each of four case file stages. **All necessary information <u>relative to violations</u> shall be obtained during the inspection, using any means deemed appropriate by the CSHO (i.e., notes, audio/videotapes, photographs, and employer records).**

      b. <u>Solicitor Coordination</u>. Consultation in accordance with regional procedures, including Solicitor procedures, shall be considered when the inspection or investigation could involve important, novel or complex litigation or when consultation is necessary in the CSHO or Area Director's professional judgment. If consultation is deemed necessary, such consultation shall be conducted at the earliest stage possible of the investigation.

   2. <u>Case File Stages</u>. The following paragraphs indicate what documentation is required for each of the four case file stages.

      NOTE: The difference between Stage III and Stage IV is one of format and organization only. A Stage III case file is not understood as involving a lesser degree of documentation.

      a. **Stage I**.

      No on-site inspection conducted ---

        o OSHA-1 or equivalent, and brief statement expanding upon the reason for not conducting the inspection.

        o If refusal of entry, information necessary to secure a warrant (see Chapter II, A.2.c.).

        o Complainant/referral response, if complaint/referral inspection.

A. 2. b. **Stage II**.

In-compliance inspection ---

o   OSHA-1 or equivalent.

o   OSHA-1A or pertinent information (see B.1. of this chapter).

o   Records obtained during the inspection, based on the CSHO's professional judgment as to what should be obtained.

> **NOTE:** The CSHO need not document that a condition was in compliance beyond a general statement that no conditions were observed in violation of any standard.

o   Complainant/referral response, if complaint/referral inspection.

c. **Stage III**.

Inspection conducted, citations to be issued ---

o   OSHA-1 or equivalent.

o   OSHA-1A or equivalent (see B.1. of this chapter).

o   Records obtained during the inspection which, based on the CSHO's professional judgment, are necessary to support the violations.

o   OSHA-1B forms or the equivalent with the following included:
Inspection #
Instances on page (a,b,/)
Type of violation (S,W,R,O,FTA)
Citation number and item number
Number exposed
REC
Abatement Period
SAVE, AVD, and/or standard reference
Photo/video location
Severity Rating (H,M,L) and brief justification
Probability Rating (G,L) and brief justification
GBP and multiplier if applicable
% reduction (adjustment)
Proposed penalty

> **NOTE:** Information in relation to exposed employees shall be documented on the OSHA-1B, or referenced on the OSHA-1B as to the specific location of this information.

o   Complainant/referral response, if complaint/referral inspection.

A.   2.   d.   **Stage IV**.

Citations are contested ---

o   CSHO's will determine after consultation with the Solicitor if the documentation obtained during the inspection needs to be transferred to a different format or location within the file (e.g., transfer of video/audio information to a written format). The information will then be transferred to the appropriate areas as needed. Items which may be considered include transfer of exposed employee information, instance description, employer knowledge, employer's affirmative defenses, employer/employee comments, and other employer information to the OSHA-1B or equivalent.

B.   **Specific Forms**.

1.   **Narrative, Form OSHA-1A**.

a.   **General**. The OSHA-1A Form, or its equivalent, shall be used to record information relative to the following, at a minimum:

ITEM: Establishment Name.

ITEM: Inspection Number.

ITEM: Additional Citation Mailing Addresses.

ITEM: Names and Addresses of all Organized Employee Groups.

ITEM: Names, Addresses, and Phone Numbers of Authorized Representatives of Employees.

ITEM: Employer Representatives Contacted and extent of their participation in the inspection.

ITEM: Comment on S&H program to the extent necessary, based on CSHO's professional judgment, including penalty reduction justifications for good faith.

ITEM: Document whether closing conference was held, describe any unusual circumstances.

ITEM: Additional Comments (CSHO's shall use their professional judgment to determine if any additional information shall be added to the case file.)

B.   1.   b.   Specific. The following information may be located on the OSHA-1A Form or referenced on the OSHA-1A as to the specific location of this information:

ITEM: Names, Addresses, and Phone Numbers of Other Persons Contacted.

ITEM: Accompanied By.

2.   Photo Mounting Worksheet, Form OSHA-89.

This worksheet may be utilized by the CSHO, if mounting is necessary. Other methods of mounting the photograph may be used, such as attaching it to the OSHA-1B. The photograph shall be annotated "trade secret," if applicable.

3.   Inspection Case File Activity Diary Insert.

The Inspection Case File Activity Diary is designed to provide a ready record and summary of all actions relating to a case. The diary sheet will be used to document important events related to the case, especially those not found elsewhere in the case file.

C.   Violations.

1.   Basis of Violations.

a.   Standards and Regulations. Section 5(a)(2) of the Occupational Safety and Health Act states that each employer has a responsibility to comply with the occupational safety and health standards promulgated under the Act. The specific standards and regulations are found in Title 29 Code of Federal Regulations (CFR) 1900 series. Subparts A and B of 29 CFR 1910 specifically establish the source of all the standards which are the basis of violations.

NOTE: The most specific subdivision of the standard shall be used for citing violations.

(1)   Definition and Application of Universal Standards (Horizontal) and Specific Industry Standards (Vertical). Specific Industry standards are

those standards which apply to a particular industry or to particular operations, practices, conditions, processes, means, methods, equipment or installations. Universal standards are those standards which apply when a condition is not covered by a specific industry standard. Within both universal and specific industry standards there are general standards and specific standards.

C. 1. a. (1) (a) When a hazard in a particular industry is covered by both a specific industry (e.g., 29 CFR Part 1915) standard and a universal (e.g., 29 CFR Part 1910) standard, the specific industry standard shall take precedence. **This is true even if the universal standard is more stringent.**

(b) When determining whether a universal or a specific industry standard is applicable to a work situation, the CSHO shall focus attention on the activity in which the employer is engaged at the establishment being inspected rather than the nature of the employer's general business.

(2) Variances. The employer's requirement to comply with a standard may be modified through granting of a variance, as outlined in Section 6 of the Act.

(a) An employer will not be subject to citation if the observed condition is in compliance with either the variance or the standard.

(b) In the event that the employer is not in compliance with the requirements of the variance, a violation of the standard shall be cited with a reference in the citation to the variance provision that has not been met.

b. Employee Exposure.

(1) Definition of Employee. Whether or not exposed persons are employees of an employer depends on several factors, the most important of which is who controls the manner in which the employees perform their assigned work. The question of who pays these employees may not be the determining factor. Determining the employer of an exposed person may be a very complex question, in which case the Area Director may seek the advice of the Regional Solicitor.

(2) Proximity to the Hazard. The proximity of the workers to the point of danger of the operation shall be documented.

C. 1. b. (3) Observed Exposure. Employee exposure is established if the CSHO witnesses, observes, or monitors exposure of an employee to the hazardous or suspected hazardous condition during work or work-related activities. Where a standard requires engineering or administrative controls (including work practice controls), employee exposure shall be cited regardless of the use of personal protective equipment.

(4) Unobserved Exposure. Where employee exposure is not observed, witnessed, or monitored by the CSHO, employee exposure is established if it is determined through witness statements or other evidence that exposure to a hazardous condition has occurred, continues to occur, or could recur.

(a) In fatality/catastrophe (or other "accident") investigations, employee exposure is established if the CSHO determines, through written statements or other evidence, that exposure to a hazardous condition occurred at the time of the accident.

(b) In other circumstances, based on the CSHO's professional judgment and determination, exposure to hazardous conditions has occurred in the past, and such exposure may serve as the basis for a violation when employee exposure has occurred in the previous six months.

(5) Potential Exposure. A citation may be issued when the possibility exists that an employee could be exposed to a hazardous condition because of work patterns, past circumstances, or anticipated work requirements, and it is reasonably predictable that employee exposure could occur, such as:

(a) The hazardous condition is an integral part of an employer's recurring operations, but the employer has not established a policy or program to ensure that exposure to the hazardous condition will not recur; or

(b) The employer has not taken steps to prevent access to unsafe machinery or equipment which employees may have reason to use.

2. Types of Violations.

a. Other-Than-Serious Violations. This type of violation shall be cited in situations where the most serious injury or illness that would be likely to result from a hazardous condition cannot reasonably be predicted to cause death or serious physical harm to exposed employees but does have a direct and immediate relationship to their safety and health.

C. 2. b. <u>Serious Violations</u>.

(1) Section 17(k) of the Act provides ". . . a serious violation shall be deemed to exist in a place of employment if there is a substantial probability that death or serious physical harm could result from a condition which exists, or from one or more practices, means, methods, operations, or processes which have been adopted or are in use, in such place of employment unless the employer did not, and could not with the exercise of reasonable diligence, know of the presence of the violation."

(2) The CSHO shall consider four elements to determine if a violation is serious.

(a) <u>Step 1</u>. The **types of accident** or health hazard exposure which the violated standard or the general duty clause is designed to prevent.

(b) <u>Step 2</u>. The most serious **injury or illness** which could reasonably be expected to result from the type of accident or health hazard exposure identified in Step 1.

(c) <u>Step 3</u>. Whether the results of the injury or illness identified in Step 2 could **include death or serious physical harm**. Serious physical harm is defined as:

<u>1</u> Impairment of the body in which part of the body is made **functionally useless** or is **substantially reduced in efficiency** on or off the job. Such impairment may be permanent or temporary, chronic or acute. Injuries involving such impairment would usually require treatment by a medical doctor.

<u>2</u> Illnesses that could shorten life or significantly reduce physical or mental efficiency by inhibiting the normal function of a part of the body.

(d) <u>Step 4</u>. Whether the **employer knew**, or with the exercise of reasonable diligence, could have known of the presence of the hazardous condition.

<u>1</u> In this regard, the supervisor represents the employer and a supervisor's knowledge of the hazardous condition amounts to employer knowledge.

<u>2</u> In cases where the employer may contend that the supervisor's own conduct constitutes an isolated event of employee misconduct, the CSHO shall attempt to determine the extent to

which the supervisor was trained and supervised so as to prevent such conduct, and how the employer enforces the rule.

<u>3</u>   If, after reasonable attempts to do so, it cannot be determined that the employer has actual knowledge of the hazardous condition, the knowledge requirement is met if the CSHO is satisfied that the employer could have known through the exercise of reasonable diligence. As a general rule, if the CSHO was able to discover a hazardous condition, and the condition was not transitory in nature, it can be presumed that the employer could have discovered the same condition through the exercise of reasonable diligence.

C. 2. c.   <u>Violations of the General Duty Clause</u>. Section 5(a)(1) of the Act requires that "Each employer shall furnish to each of his (sic) employees employment and a place of employment which are free from recognized hazards that are causing or are likely to cause death or serious physical harm to his (sic) employees." The general duty provisions shall be used only where there is no standard that applies to the particular hazard involved, as outlined in 29 CFR §1910.5(f).

(1)   <u>Evaluation of Potential Section 5(a)(1) Situations</u>. In general, Review Commission and court precedent has established that the following elements are necessary to prove a violation of the general duty clause:

(a)   The employer failed to keep the workplace free of a hazard to which employees of that employer were exposed;

(b)   The hazard was recognized;

(c)   The hazard was causing or was likely to cause death or serious physical harm; and

(d)   There was a feasible and useful method to correct the hazard.

(2)   <u>Discussion of Section 5(a)(1) Elements</u>. The above four elements of a Section 5(a)(1) violation are discussed in greater detail as follows:

(a)   <u>A Hazard to Which Employees Were Exposed</u>. A general duty citation must involve both a serious hazard and exposure of employees.

C. 2. c. (2) (a) <u>1</u> <u>Hazard</u>. A hazard is a danger which threatens physical harm to employees.

       <u>a</u> <u>Not the Lack of a Particular Abatement Method</u>. In the past some Section 5(a)(1) citations have incorrectly alleged that the violation is the failure to implement certain precautions, corrective measures or other abatement steps rather than the failure to prevent or remove the particular hazard. It must be emphasized that Section 5(a)(1) does not mandate a particular abatement measure but only requires an employer to render the workplace free of certain hazards by any feasible and effective means which the employer wishes to utilize.

       EXAMPLE: In a hazardous situation involving high pressure gas where the employer has failed to train employees properly, has not installed the proper high pressure equipment, and has improperly installed the equipment that is in place, there are three abatement measures which the employer failed to take; there is only one hazard (that is, exposure to the hazard of explosion due to the presence of high pressure gas) and hence only one general duty clause citation.

       <u>b</u> <u>The Hazard Is Not a Particular Accident</u>. The occurrence of an accident does not necessarily mean that the employer has violated Section 5(a)(1) although the accident may be evidence of a hazard. In some cases a Section 5(a)(1) violation may be unrelated to the accident. Although accident facts may be relevant and shall be gathered, the citation shall address the hazard in the workplace, not the particular facts of the accident.

       EXAMPLE: A fire occurred in a workplace where flammable materials were present. No employee was injured by the fire itself but an employee, disregarding the clear instructions of his/her supervisor to use an available exit, jumped out of a window and broke a leg. The danger of fire due to the presence of flammable materials may be a recognized hazard causing or likely to cause death or serious physical harm, but the action of the employee may be an instance of unpreventable employee misconduct. The citation should deal with the fire hazard, not with the accident involving the employee who broke his/her leg.

C. 2. c. (2) (a) <u>1</u> <u>c</u>   <u>The Hazard Must Be Reasonably Foreseeable</u>. The hazard for which a citation is issued must be reasonably foreseeable.

    i. All the factors which could cause a hazard need not be present in the same place at the same time in order to prove foreseeability of the hazard; e.g., an explosion need not be imminent.

        EXAMPLE: If combustible gas and oxygen are present in sufficient quantities in a confined area to cause an explosion if ignited but no ignition source is present or could be present, no Section 5(a)(1) violation would exist. If an ignition source is available at the workplace and the employer has not taken sufficient safety pre-cautions to preclude its use in the confined area, then a foreseeable hazard may exist.

    ii. It is necessary to establish the reasonable foreseeability of the general workplace hazard, rather than the partic-ular hazard which led to the accident.

        EXAMPLE: A titanium dust fire may have spread from one room to another only because an open can of gasoline was in the second room. An employee who usually worked in both rooms was burned in the second room from the gasoline. The presence of gasoline in the second room may be a rare occurrence. It is not necessary to prove that a fire in both rooms was reasonably foreseeable. It is necessary only to prove that the fire hazard, in this case due to the presence of titanium dust, was reasonably foreseeable.

    <u>2</u>   <u>The Hazard Must Affect the Cited Employer's Employees</u>. The employees exposed to the Section 5(a)(1) hazard must be the employees of the cited employer.

  (b) <u>The Hazard Must be Recognized</u>. Recognition of a hazard can be established on the basis of industry recognition, employer recogni-tion, or "common-sense" recognition. The use of common-sense as the basis for establishing recognition shall be limited to special cir-cumstances. Recognition of the hazard must be supported by satis-factory evidence and adequate documentation in the file as follows:

C. 2. c. (2) (b) <u>1</u> <u>Industry Recognition</u>. A hazard is recognized if the employer's industry recognizes it. Recognition by an industry other than the industry to which the employer belongs is generally insufficient to prove this element of a Section 5(a)(1) violation. Although evidence of recognition by the employer's specific branch within an industry is preferred, evidence that the employer's industry recognizes the hazard may be sufficient.

<u>a</u> In cases where State and local government agencies not falling under the preemption provisions of Section 4(b)(1) have codes or regulations covering hazards not addressed by OSHA standards, the Area Director shall determine whether the hazard is to be cited under Section 5(a)(1) or referred to the appropriate local agency for enforcement.

<u>b</u> Regulations of other Federal agencies or of State atomic energy agencies generally shall not be used. They raise substantial difficulties under Section 4(b)(1) of the Act, which provides that OSHA is preempted when such an agency has statutory authority to deal with the working condition in question.

<u>2</u> <u>Employer Recognition</u>. A recognized hazard can be established by evidence of actual employer knowledge. Evidence of such recognition may consist of written or oral statements made by the employer or other management or supervisory personnel during or before the OSHA inspection, or instances where employees have clearly called the hazard to the employer's attention.

<u>3</u> <u>Common-Sense Recognition</u>. If industry or employer recognition of the hazard cannot be established in accordance with (a) and (b), recognition can still be established if it is concluded that any reasonable person would have recognized the hazard. This theory of recognition shall be used only in flagrant cases.

(c) <u>The Hazard Was Causing or Was Likely to Cause Death or Serious Physical Harm</u>. This element of Section 5(a)(1) violation is identical to the elements of a serious violation, see C.2.b. of this chapter.

(d) <u>The Hazard Can Be Corrected by a Feasible and Useful Method</u>.

<u>1</u> To establish a Section 5(a)(1) violation the agency must identify a method which is feasible, available and likely to

correct the hazard. The information shall indicate that the recognized hazard, rather than a particular accident, is preventable.

C. 2. c. (2) (d) 2   If the proposed abatement method would eliminate or significantly reduce the hazard beyond whatever measures the employer may be taking, a Section 5(a)(1) citation may be issued. A citation shall not be issued merely because the agency knows of an abatement method different from that of the employer, if the agency's method would not reduce the hazard significantly more than the employer's method. It must also be noted that in some cases only a series of abatement methods will alleviate a hazard. In such a case all the abatement methods shall be mentioned.

(3)   Limitations on Use of the General Duty Clause. Section 5(a)(1) is to be used only within the guidelines given in C.2.c. of this chapter.

(a)   Section 5(a)(1) may be cited in the alternative when a standard is also cited to cover a situation where there is doubt as to whether the standard applies to the hazard.

(b)   Section 5(a)(1) violations shall not be grouped together, but may be grouped with a related violation of a specific standard.

(c)   Section 5(a)(1) shall not normally be used to impose a stricter requirement than that required by the standard. For example, if the standard provides for a permissible exposure limit (PEL) of 5 ppm, even if data establishes that a 3 ppm level is a recognized hazard, Section 5(a)(1) shall not be cited to require that the 3 ppm level be achieved unless the limits are based on different health effects. If the standard has only a time-weighted average permissible exposure level and the hazard involves exposure above a recognized ceiling level, the Area Director shall consult with the Regional Solicitor.

NOTE: An exception to this rule may apply if it can be documented that "an employer knows a particular safety or health standard is inadequate to protect his workers against the specific hazard it is intended to address." International Union, U.A.W. v. General Dynamics Land Systems Div., 815 F.2d 1570 (D.C. Cir. 1987). Such cases shall be subject to pre-citation review.

C. 2. c. (3) (d) Section 5(a)(1) shall normally not be used to require an abatement method not set forth in a specific standard. If a toxic substance standard covers engineering control requirements but not requirements for medical surveillance, Section 5(a)(1) shall not be cited to require medical surveillance.

(e) Section 5(a)(1) shall not be used to enforce "should" standards.

(f) Section 5(a)(1) shall not normally be used to cover categories of hazards exempted by a standard. If, however, the exemption is in place because the drafters of the standard (or source document) declined to deal with the exempt category for reasons other than the lack of a hazard, the general duty clause may be cited if all the necessary elements for such a citation are present.

(4) Pre-Citation Review. Section 5(a)(1) citations shall undergo a pre-citation review following established area office procedures when required by the Area Director or Assistant Area Director.

NOTE: If a standard does not apply and all criteria for issuing a Section 5(a)(1) citation are not met, but it is determined that the hazard warrants some type of notification, a letter shall be sent to the employer and the employee representative describing the hazard and suggesting corrective action.

d. Willful Violations. The following definitions and procedures apply whenever the CSHO suspects that a willful violation may exist:

(1) A willful violation exists under the Act where the evidence shows either an intentional violation of the Act or plain indifference to its requirements.

(a) The employer committed an intentional and knowing violation if:

1 An employer representative was aware of the requirements of the Act, or the existence of an applicable standard or regulation, and was also aware of a condition or practice in violation of those requirements, and did not abate the hazard.

2 An employer representative was not aware of the requirements of the Act or standards, but was aware of a comparable legal requirement (e.g., state or local law) and was also aware of a condition or practice in violation of that requirement, and did not abate the hazard.

C. 2. d. (1) (b) The employer committed a violation with plain indifference to the law where:

    <u>1</u> Higher management officials were aware of an OSHA requirement applicable to the company's business but made little or no effort to communicate the requirement to lower level supervisors and employees.

    <u>2</u> Company officials were aware of a continuing compliance problem but made little or no effort to avoid violations.

        EXAMPLE: Repeated issuance of citations addressing the same or similar conditions.

    <u>3</u> An employer representative was not aware of any legal requirement, but was aware that a condition or practice was hazardous to the safety or health of employees and made little or no effort to determine the extent of the problem or to take the corrective action. Knowledge of a hazard may be gained from such means as insurance company reports, safety committee or other internal reports, the occurrence of illnesses or injuries, media coverage, or, in some cases, complaints of employees or their representatives.

    <u>4</u> Finally, in particularly flagrant situations, willfulness can be found despite lack of knowledge of either a legal requirement or the existence of a hazard if the circumstances show that the employer would have placed no importance on such knowledge even if he or she had possessed it, or had no concern for the health or safety of employees.

(2) It is not necessary that the violation be committed with a bad purpose or an evil intent to be deemed "willful." It is sufficient that the violation was deliberate, voluntary or intentional as distinguished from inadvertent, accidental or ordinarily negligent.

(3) The CSHO shall carefully develop and record, during the inspection, all evidence available that indicates employer awareness of and the disregard for statutory obligations or of the hazardous conditions. Willfulness could exist if an employer is advised by employees or employee representatives of an alleged hazardous condition and the employer makes no reasonable effort to verify and correct the condition. Additional factors which can influence a decision as to whether violations are willful include:

C. 2. d. (3) (a) The nature of the employer's business and the knowledge regarding safety and health matters which could reasonably be expected in the industry.

(b) The precautions taken by the employer to limit the hazardous conditions.

(c) The employer's awareness of the Act and of the responsibility to provide safe and healthful working conditions.

(d) Whether similar violations and/or hazardous conditions have been brought to the attention of the employer.

(e) Whether the nature and extent of the violations disclose a **purposeful disregard** of the employer's responsibility under the Act.

(4) If the Area Office cannot determine whether to issue a citation as a willful or a repeat violation due to the raising of difficult issues of law and policy which will require the evaluation of complex factual situations, the Area Director shall normally consult with the Regional Solicitor.

e. Criminal/Willful Violations. Section 17(e) of the Act provides that: "Any employer who willfully violates any standard, rule or order promulgated pursuant to Section 6 of this Act, or of any regulations prescribed pursuant to this Act, and that violation caused death to any employee, shall, upon conviction, be punished by a fine of not more than $10,000 or by imprisonment for not more than six months, or by both; except that if the conviction is for a violation committed after a first conviction of such person, punishment shall be a fine of not more than $20,000 or by imprisonment for not more than one year, or by both."

(1) The Area Director, in coordination with the Regional Solicitor, shall carefully evaluate all willful cases involving worker deaths to determine whether they may involve criminal violations of Section 17(e) of the Act. Because the nature of the evidence available is of paramount importance in an investigation of this type, there shall be early and close liaison between the OSHA investigator, the Area Director, the Regional Administrator, and the Regional Solicitor in developing any finding which might involve a violation of Section 17(e) of the Act.

C. 2. e. (2) The following criteria shall be considered in investigating possible criminal/willful violations:

   (a) In order to establish a criminal/willful violation OSHA must prove that:

      1 The employer violated an OSHA standard. A criminal/willful violation cannot be based on violation of Section 5(a)(1).

      2 The violation was willful in nature.

      3 The violation of the standard caused the death of an employee. In order to prove that the violation of the standard caused the death of an employee, there must be evidence in the file which clearly demonstrates that the violation of the standard was the cause of or a contributing factor to an employee's death.

   (b) Although it is generally not necessary to issue "Miranda" warnings to an employer when a criminal/willful investigation is in progress, the Area Director shall seek the advice of the Regional Solicitor on this question.

   (c) Following the investigation, if the Area Director decides to recommend criminal prosecution, a memorandum containing that recommendation shall be forwarded promptly to the Regional Administrator. It shall include an evaluation of the possible criminal charges, taking into consideration the greater burden of proof which requires that the Government's case be proven beyond a reasonable doubt. In addition, if the correction of the hazardous condition appears to be an issue, this shall be noted in the transmittal memorandum because in most cases the prosecution of a criminal/willful case delays the affirmance of the civil citation and its correction requirements.

   (d) The Area Director shall normally issue a civil citation in accordance with current procedures even if the citation involves allegations under consideration for criminal prosecution. The Regional Administrator shall be notified of such cases, and they shall be forwarded to the Regional Solicitor as soon as practicable for possible referral to the U.S. Department of Justice.

   (3) When a willful violation is related to a fatality, the Area Director shall ensure the case file contains succinct documentation regarding the

decision **not** to make a criminal referral. The documentation should indicate which elements of a criminal violation make the case unsuitable for criminal referral.

C. 2. f. <u>Repeated Violations</u>. An employer may be cited for a repeated violation if that employer has been cited previously for a **substantially similar condition** and the citation has become a final order.

(1) <u>Identical Standard</u>. Generally, similar conditions can be demonstrated by showing that in both situations the identical standard was violated.

> **EXCEPTION:** Previously a citation was issued for a violation of 29 CFR §1910.132(a) for not requiring the use of safety-toe footwear for employees. A recent inspection of the same establishment revealed a violation of 29 CFR §1910.132(a) for not requiring the use of head protection (hard hats). Although the same standard was involved, the hazardous conditions found were not substantially similar and therefore a repeated violation would not be appropriate.

(2) <u>Different Standards</u>. In some circumstances, similar conditions can be demonstrated when different standards are violated. Although there may be different standards involved, the hazardous conditions found could be substantially similar and therefore a repeated violation would be appropriate.

(3) <u>Time Limitations</u>. Although there are no statutory limitations upon the length of time that a citation may serve as a basis for a repeated violation, the following policy shall be used in order to ensure uniformity.

(a) A citation will be issued as a repeated violation if:

<u>1</u> The citation is issued within 3 years of the final order of the previous citation, or,

<u>2</u> The citation is issued within 3 years of the final abatement date of that citation, whichever is later.

(b) When a violation is found during an inspection, and a repeated citation has been issued for a substantially similar condition which meets the above time limitations, the violation may be classified as a second instance repeated violation with a corresponding increase in penalty (see Chapter IV, C.2.l.).

C. 2. f. (3) (c) For any further repetition, the Area Director shall be consulted for guidance.

(4) Obtaining Inspection History. For purposes of determining whether a violation is repeated, the following criteria shall apply:

(a) High Gravity Serious Violations. When high gravity serious violations are to be cited, the Area Director shall obtain a history of citations previously issued to this employer at all of its identified establishments, nationwide, (Federal enforcement only) within the same two-digit SIC code. If these violations have been previously cited within the time limitations described in C.2.f.(3), above, and have become a final order of the Review Commission, a repeated citation may be issued. Under special circumstances, the Area Director, in consultation with the Regional Solicitor, may also issue citations for repeated violations without regard for the SIC code.

(b) Violations of Lesser Gravity. When violations of lesser gravity than high gravity serious are to be cited, Agency policy is to encourage the Area Director to obtain a national inspection history whenever the circumstances of the current inspection will result in a large number of serious, repeat, or willful citations. This is particularly so if the employer is known to have establishments nationwide and if significant citations have been issued against the employer in other areas, or at other mobile worksites.

(c) Geographical Limitations. Where a national inspection history has **not** been obtained, the following criteria regarding geographical limitations shall apply:

1 Multifacility Employer. A multifacility employer shall be cited for a repeated violation if the violation recurred at any worksite within the same OSHA Area Office jurisdiction.

EXAMPLE: Where the construction site extends over a large area and/or the scope of the job is unclear (such as road building), that portion of the workplace specified in the employer's contract which falls within the Area Office jurisdiction is the establishment. If an employer has several worksites within the same Area Office jurisdiction, a citation of a violation at Site A will serve as the basis for a repeated citation in Area B.

2 Longshoring Establishment. A longshoring establishment will encompass all longshoring activities of a single stevedore within any single port area. Longshoring employers are subject

to repeated violation citations based on prior violations occurring anywhere. Other maritime employers covered by OSHA standards (e.g., shipbuilding, ship repairing) are multifacility employers as defined in a., above.

C. 2. f. (5) Repeated vs. Willful. Repeated violations differ from willful violations in that they may result from an inadvertent, accidental or ordinarily negligent act. Where a repeated violation may also meet the criteria for willful but not clearly so, a citation for a repeated violation shall normally be issued.

(6) Repeated vs. Failure to Abate. A failure to abate situation exists when an item of equipment or condition previously cited has never been brought into compliance and is noted at a later inspection. If, however, the violation was not continuous (i.e., if it had been corrected and then reoccurred), the subsequent occurrence is a repeated violation.

(7) Alleged Violation Description (AVD). If a repeated citation is issued, the CSHO must ensure that the cited employer is fully informed of the previous violations serving as a basis for the repeated citation, by notation in the AVD portion of the citation, using the following or similar language:

THE (COMPANY NAME) WAS PREVIOUSLY CITED FOR A VIOLATION OF THIS OCCUPATIONAL SAFETY AND HEALTH STANDARD OR ITS EQUIVALENT STANDARD (NAME PREVIOUSLY CITED STANDARD) WHICH WAS CONTAINED IN OSHA INSPECTION NUMBER_____, CITATION NUMBER_____, ITEM NUMBER_____, ISSUED ON (DATE), WITH RESPECT TO A WORKPLACE LOCATED AT_____.

g. De Minimis Violations. De Minimis violations are violations of standards which have no direct or immediate relationship to safety or health and shall not be included in citations. An OSHA-1B/1BIH is no longer required to be completed for De Minimis violations. The employer should be verbally notified of the violation and the CSHO should note it in the inspection case file. The criteria for finding a de minimis violation are as follows:

(1) An employer complies with the clear intent of the standard but deviates from its particular requirements in a manner that has no direct or immediate relationship to employee safety or health. These deviations may involve distance specifications, construction material requirements, use of incorrect color, minor variations from recordkeeping, testing, or inspection regulations, or the like.

EXAMPLE #1: 29 CFR §1910.27(b)(1)(ii) allows 12 inches (30 centimeters) as the maximum distance between ladder rungs. Where the rungs are 13 inches (33 centimeters) apart, the condition is de minimis.

EXAMPLE #2: 29 CFR §1910.28(a)(3) requires guarding on all open sides of scaffolds. Where employees are tied off with safety belts in lieu of guarding, often the intent of the standard will be met, and the absence of guarding may be de minimis.

EXAMPLE #3: 29 CFR §1910.217(e)(1)(ii) requires that mechanical power presses be inspected and tested at least weekly. If the machinery is seldom used, inspection and testing prior to each use is adequate to meet the intent of the standard.

(2) An employer complies with a proposed standard or amendment or a consensus standard rather than with the standard in effect at the time of the inspection and the employer's action clearly provides equal or greater employee protection or the employer complies with a written interpretation issued by the OSHA Regional or National Office.

(3) An employer's workplace is at the "state of the art" which is technically beyond the requirements of the applicable standard and provides equivalent or more effective employee safety or health protection.

C. 3. <u>Health Standard Violations</u>.

    a. <u>Citation of Ventilation Standards</u>. In cases where a citation of a ventilation standard may be appropriate, consideration shall be given to standards intended to control exposure to recognized hazardous levels of air contaminants, to prevent fire or explosions, or to regulate operations which may involve confined space or specific hazardous conditions. In applying these standards, the following guidelines shall be observed:

        (1) <u>Health-Related Ventilation Standards</u>. An employer is considered in compliance with a health-related airflow ventilation standard when the employee exposure does not exceed appropriate airborne contaminant standards; e.g., the PELs prescribed in 29 CFR 1910.1000.

            (a) Where an over-exposure to an airborne contaminant is detected, the appropriate air contaminant engineering control requirement shall be cited; e.g., 29 CFR 1910.1000(e). In no case shall citations of this standard be issued for the purpose of requiring specific volumes of air to ventilate such exposures.

(b) Other requirements contained in health-related ventilation standards shall be evaluated without regard to the concentration of airborne contaminants. Where a specific standard has been violated <u>and</u> an actual or potential hazard has been documented, a citation shall be issued.

C. 3. a. (2) <u>Fire- and Explosion-Related Ventilation Standards</u>. Although they are not technically health violations, the following guidelines shall be observed when citing fire- and explosion-related ventilation standards:

    (a) <u>Adequate Ventilation</u>. In the application of fire- and explosion-related ventilation standards, OSHA considers that an operation has **adequate** ventilation when both of the following criteria are met:

        <u>1</u> The requirement of the specific standard has been met.

        <u>2</u> The concentration of flammable vapors is 25 percent or less of the lower explosive limit (LEL).

        EXCEPTION: Certain standards specify violations when 10 percent of the LEL is exceeded. These standards are found in maritime and construction exposures.

    (b) <u>Citation Policy</u>. If 25 percent (10 percent when specified for maritime or construction operations) of the LEL has been exceeded and:

        <u>1</u> The standard requirements have not been met, the standard violation normally shall be cited as serious.

        <u>2</u> There is no applicable specific ventilation standard, Section 5(a)(1) of the Act shall be cited in accordance with the guidelines given in C.2.c. of this chapter.

  b. <u>Violations of the Noise Standard</u>. Current enforcement policy regarding 29 CFR 1910.95(b)(1) allows employers to rely on personal protective equipment and a hearing conservation program rather than engineering and/or administrative controls when hearing protectors will effectively attenuate the noise to which the employee is exposed to acceptable levels as specified in Tables G-16 or G-16a of the standard.

    (1) Citations for violations of 29 CFR 1910.95(b)(1) shall be issued when engineering and/or administrative controls are feasible, both technically and economically; and

C. 3. b. (1) (a) Employee exposure levels are so high that hearing protectors alone may not reliably reduce noise levels received by the employee's ear to the levels specified in Tables G-16 or G-16a of the standard. Given the present state of the art, hearing protectors which offer the greatest attenuation may not reliably be used when employee exposure levels border on 100 dBA (See OSHA Instruction CPL 2-2.35A, Appendix.); or

(b) The costs of engineering and/or administrative controls are less than the cost of an effective hearing conservation program.

(2) A control is not reasonably necessary when an employer has an ongoing hearing conservation program and the results of audiometric testing indicate that existing controls and hearing protectors are adequately protecting employees. (In making this decision such factors as the exposure levels in question, the number of employees tested, and the duration of the testing program shall be taken into consideration.)

(3) When employee noise exposures are less than 100 dBA but the employer does not have an ongoing hearing conservation program or the results of audiometric testing indicate that the employer's existing program is not working, the CSHO shall consider whether:

(a) Reliance on an effective hearing conservation program would be less costly than engineering and/or administrative controls.

(b) An effective hearing conservation program can be established or improvements can be made in an existing hearing conservation program which could bring the employer into compliance with Tables G-16 or G-16a.

(c) Engineering and/or administrative controls are both technically and economically feasible.

(4) If noise levels received by the employee's ear can be reduced to the levels specified in Tables G-16 or G-16a by means of hearing protectors and an effective hearing conservation program, citations under the hearing conservation standard shall normally be issued rather than citations requiring engineering controls. If improvements in the hearing conservation program cannot be made or, if made, cannot be expected to reduce exposure sufficiently and feasible controls exist, a citation under 1910.95(b)(1) shall normally be issued.

(5) When hearing protection is required but not used and employee exposure exceeds the limits of Table G-16, 29 CFR 1910.95(i)(2)(i) shall be

cited and classified as serious (see (8), below) whether or not the employer has instituted a hearing conservation program. 29 CFR 1910.95(a) shall no longer be cited except in the case of the oil and gas drilling industry.

NOTE: Citations of 29 CFR 1910.95(i)(2)(ii)(b) shall also be classified as serious.

C. 3. b. (6) If an employer has instituted a hearing conservation program and a violation of the hearing conservation amendment (other than 1910.95 (i)(2)(i) or (i)(2)(ii)(b)) is found, a citation shall be issued if employee noise exposures equal or exceed an 8-hour time-weighted average of 85 dB.

(7) If the employer has not instituted a hearing conservation program and employee noise exposures equal or exceed an 8-hour time-weighted average of 85 dB, a citation for 1910.95(c) only shall be issued.

(8) Violations of 1910.95(i)(2)(i) from the hearing conservation amendment may be grouped with violations of 29 CFR 1910.95(b)(1) and classified as serious when an employee is exposed to noise levels above the limits of Table G-I6 and:

(a) Hearing protection is not utilized or is not adequate to prevent over-exposure to an employee; or

(b) There is evidence of hearing loss which could reasonably be considered:

<u>1</u>   To be work-related, and

<u>2</u>   To have been preventable, at least to some degree, if the employer had been in compliance with the cited provisions.

(9) When an employee is overexposed but effective hearing protection is being provided and used, an effective hearing conservation program has been implemented and no feasible engineering or administrative controls exist, a citation shall not be issued.

c. <u>Violations of the Respirator Standard</u>. When considering a citation for respirator violations, the following guidelines shall be observed:

(1) <u>In Situations Where Overexposure Does Not Occur</u>. Where an overexposure has not been established:

C. 3. c. (1) (a) But an improper type of respirator is being used (e.g., a dust respirator being used to reduce exposure to organic vapors), a citation under 29 CFR 1910.134(b)(2) shall be issued, provided the CSHO documents that an overexposure is possible.

(b) And one or more of the other requirements of 29 CFR 1910.134 is not being met; e.g., an unapproved respirator is being used to reduce exposure to toxic dusts, generally a de minimis violation shall be recorded in accordance with OSHA procedures. (Note that this policy does not include emergency use respirators.) The CSHO shall advise the employer of the elements of a good respirator program as required under 29 CFR 1910.134.

(c) In exceptional circumstances a citation may be warranted if an adverse health condition due to the respirator itself could be supported and documented. Examples may include a dirty respirator that is causing dermatitis, a worker's health being jeopardized by wearing a respirator due to an inadequately evaluated medical condition or a significant ingestion hazard created by an improperly cleaned respirator.

(2) In Situations Where Overexposure Does Occur. In cases where an overexposure to an air contaminant has been established, the following principles apply to citations of 1910.134:

(a) 29 CFR 1910.134(a)(2) is the general section requiring employers to provide respirators ". . . when such equipment is necessary to protect the health of the employee" and requiring the establishment and maintenance of a respiratory protection program which meets the requirements outlined in 29 CFR 1910.134(b). Thus, if no respiratory program at all has been established, 1910.134(a)(2) alone shall be cited; if a program has been established and some, but not all, of the requirements under 1910.134(b) are being met, the specific standards under 1910.134(b) that are applicable shall be cited.

(b) An acceptable respiratory protection program includes all of the elements of 29 CFR 1910.134; however, the standard is structured such that essentially the same requirement is often specified in more than one section. In these cases, the section which most adequately describes the violation shall be cited.

C. 3. d. <u>Additive and Synergistic Effects</u>.

(1) Substances which have a known additive effect and, therefore, result in a greater probability/severity of risk when found in combination shall be evaluated using the formula found in 29 CFR § 1910.1000(d)(2). The use of this formula requires that the exposures have an additive effect on the same body organ or system.

(2) If the CSHO suspects that synergistic effects are possible, it shall be brought to the attention of the supervisor, who shall refer the question to the Regional Administrator. If it is decided that there is a synergistic effect of the substances found together, the violations shall be grouped, when appropriate, for purposes of increasing the violation classification severity and/or the penalty.

e. <u>Absorption and Ingestion Hazards</u>. The following guidelines apply when citing absorption and ingestion violations. Such citations do <u>not</u> depend on measurements of airborne concentrations, but shall normally be supported by wipe sampling.

(1) Citations under 29 CFR 1910.132, 1910.141 and/or Section 5(a)(1) may be issued when there is reasonable probability that employees will be exposed to these hazards.

(2) Where, for any substance, a serious hazard is determined to exist due to the potential of ingestion or absorption of the substance for reasons other than the consumption of contaminated food or drink (e.g., smoking materials contaminated with the toxic substance), a serious citation shall be considered under Section 5(a)(1) of the Act.

f. <u>Biological Monitoring</u>. If the employer has been conducting biological monitoring, the CSHO shall evaluate the results of such testing. The results may assist in determining whether a significant quantity of the toxic material is being ingested or absorbed through the skin.

4. <u>Writing Citations</u>.

a. <u>General</u>. Section 9 of the Act controls the writing of citations.

(1) <u>Section 9(a)</u>. ". . . the Secretary or his authorized representative . . . shall with reasonable promptness issue a citation to the employer." To facilitate the prompt issuance of citations, the Area Director may issue citations which are unrelated to health inspection air sampling, prior to receipt of sampling results.

C. 4. a. (2) <u>Section 9(c)</u>. "No citation may be issued . . . after the expiration of six months following the occurrence of any violation." Accordingly, a citation shall not be issued where any violation alleged therein last occurred 6 months or more prior to the date on which the citation is actually signed and dated. Where the actions or omissions of the employer concealed the existence of the violation, the time limitation is suspended until such time that OSHA learns or could have learned of the violation. The Regional Solicitor shall be consulted in such cases.

b. <u>Alternative Standards</u>.

(1) In rare cases, the same factual situation may present a possible violation of more than one standard. For example, the facts which support a violation of 29 CFR §1910.28(a)(1) may also support a violation of §1910.132(a) if no scaffolding is provided when it should be and the use of safety belts is not required by the employer.

(2) Where it appears that more than one standard is applicable to a given factual situation and that compliance with any of the applicable standards would effectively eliminate the hazard, it is permissible to cite alternative standards using the words "in the alternative." A reference in the citation to each of the standards involved shall be accompanied by a separate Alleged Violation Description (AVD) which clearly alleges all of the necessary elements of a violation of that standard. Only one penalty shall be proposed for the violative condition.

5. <u>Combining and Grouping of Violations</u>.

a. <u>Combining</u>. Violations of a single standard having the same classification found during the inspection of an establishment or worksite generally shall be combined into one alleged citation item. Different options of the same standard shall normally also be combined. Each instance of the violation shall be separately set out within that item of the citation. Other-than-serious violations of a standard may be combined with serious violations of the same standard when appropriate.

NOTE: Except for standards which deal with multiple hazards (e.g., Tables Z-1, Z-2 and Z-3 cited under 29 CFR §1910.1000(a), (b), or (c)), the same standard may not be cited more than once on a single citation. The same standard may be cited on different citations on the same inspection, however.

b. <u>Grouping</u>. When a source of a <u>hazard</u> is identified which involves interrelated violations of different standards, the violations may be grouped into a single item. The following situations normally call for grouping violations:

C. 5. b. (1) Grouping Related Violations. When the CSHO believes that violations classified either as serious or as other-than-serious are so closely related as to constitute a single hazardous condition.

(2) Grouping Other-Than-Serious Violations Where Grouping Results in a Serious Violation. When two or more individual violations are found which, if considered individually represent other-than-serious violations, but if grouped create a substantial probability of death or serious physical harm.

(3) Where Grouping Results in Higher Gravity Other-Than-Serious Violation. Where the CSHO finds during the course of the inspection that a number of other-than-serious violations are present in the same piece of equipment which, considered in relation to each other affect the overall gravity of possible injury resulting from an accident involving the combined violations.

(4) Violations of Posting and Recordkeeping Requirements. Violations of the posting and recordkeeping requirements which involve the same document; e.g., OSHA-200 Form was not posted or maintained. (See Chapter IV, C.2.n. for penalty amounts.)

(5) Penalties for Grouped Violations. If penalties are to be proposed for grouped violations, the penalty shall be written across from the first violation item appearing on the OSHA-2.

c. When Not to Group. Times when grouping is normally inappropriate.

(1) Multiple Inspections. Violations discovered in multiple inspections of a single establishment or worksite may not be grouped. An inspection in the same establishment or at the same worksite shall be considered a single inspection even if it continues for a period of more than one day or is discontinued with the intention of resuming it after a short period of time if only one OSHA-1 is completed.

(2) Separate Establishments of the Same Employer. Where inspections are conducted, either at the same time or different times, at two establishments of the same employer and instances of the same violation are discovered during each inspection, the employer shall be issued separate citations for each establishment. The violations shall not be grouped.

(3) General Duty Clause Violations. Because Section 5(a)(1) of the Act is cited so as to cover all aspects of a serious hazard for which no standard exists, no grouping of separate Section 5(a)(1) violations is

permitted. This provision, however, does not prohibit grouping a Section 5(a)(1) violation with a related violation of a specific standard.

(4) Egregious Violations. Violations which are proposed as violation-by-violation citations shall **not** normally be combined or grouped. (See OSHA Instruction CPL 2.80.)

C. 6. Multiemployer Worksites. On multiemployer worksites, both construction and non-construction, citations normally shall be issued to employers whose employees are exposed to hazards (the exposing employer).

a. Additionally, the following employers normally shall be cited, whether or not their own employees are exposed, but see C.2.c.(2)(a)2 of this chapter for Section 5(a)(1) violation guidance:

(1) The employer who actually creates the hazard (the creating employer);

(2) The employer who is responsible, by contract or through actual practice, for safety and health conditions on the worksite; i.e., the employer who has the authority for ensuring that the hazardous condition is corrected (the controlling employer);

(3) The employer who has the responsibility for actually correcting the hazard (the correcting employer).

b. Prior to issuing citations to an exposing employer, it must first be determined whether the available facts indicate that employer has a legitimate defense to the citation, as set forth below:

(1) The employer did not create the hazard;

(2) The employer did not have the responsibility or the authority to have the hazard corrected;

(3) The employer did not have the ability to correct or remove the hazard;

(4) The employer can demonstrate that the creating, the controlling and/or the correcting employers, as appropriate, have been specifically notified of the hazards to which his/her employees are exposed;

(5) The employer has instructed his/her employees to recognize the hazard and, where necessary, informed them how to avoid the dangers associated with it.

(a) Where feasible, an exposing employer must have taken appropriate alternative means of protecting employees from the hazard.

(b) When extreme circumstances justify it, the exposing employer shall have removed his/her employees from the job to avoid citation.

C. 6. c. If an exposing employer meets all these defenses, that employer shall not be cited. If all employers on a worksite with employees exposed to a hazard meet these conditions, then the citation shall be issued only to the employers who are responsible for creating the hazard and/or who are in the best position to correct the hazard or to ensure its correction. In such circumstances the controlling employer and/or the hazard-creating employer shall be cited even though no employees of those employers are exposed to the violative condition. Penalties for such citations shall be appropriately calculated, using the exposed employees of all employers as the number of employees for probability assessment.

7. Employer/Employee Responsibilities.

a. Section 5(b) of the Act states: "Each employee shall comply with occupational safety and health standards and all rules, regulations, and orders issued pursuant to the Act which are applicable to his own actions and conduct." The Act does not provide for the issuance of citations or the proposal of penalties against employees. Employers are responsible for employee compliance with the standards.

b. In cases where the CSHO determines that employees are systematically refusing to comply with a standard applicable to their own actions and conduct, the matter shall be referred to the Area Director who shall consult with the Regional Administrator.

c. Under no circumstances is the CSHO to become involved in an onsite dispute involving labor-management issues or interpretation of collective-bargaining agreements. The CSHO is expected to obtain enough information to understand whether the employer is using all appropriate authority to ensure compliance with the Act. Concerted refusals to comply will not bar the issuance of an appropriate citation where the employer has failed to exercise full authority to the maximum extent reasonable, including discipline and discharge.

8. Affirmative Defenses.

a. Definition. An affirmative defense is any matter which, if established by the employer, will excuse the employer from a violation which has otherwise been proved by the CSHO.

b. Burden of Proof. Although affirmative defenses must be proved by the employer at the time of the hearing, OSHA must be prepared to respond whenever the employer is likely to raise or actually does raise an argument supporting such a defense. The CSHO, therefore, shall keep in mind the potential affirmative defenses that the employer may make and attempt to gather contrary evidence when a statement made during the inspection fairly raises a defense. The CSHO should bring the documentation of the hazards and facts related to possible affirmative defenses to the attention of the Assistant Area Director. Where it appears that each and every element of an affirmative defense is present, the Area Director may decide that a citation is not warranted.

C. 8. c. Explanations. The following are explanations of the more common affirmative defenses with which the CSHO shall become familiar. There are other affirmative defenses besides these, but they are less frequently raised or are such that the facts which can be gathered during the inspection are minimal.

(1) Unpreventable Employee Misconduct or "Isolated Event". The violative condition was:

(a) Unknown to the employer; and

(b) In violation of an adequate work rule which was effectively communicated and uniformly enforced.

EXAMPLE: An unguarded table saw is observed. The saw, however, has a guard which is reattached while the CSHO watches. Facts which the CSHO shall document may include: Who removed the guard and why? Did the employer know that the guard had been removed? How long or how often had the saw been used without guards? Did the employer have a work rule that the saw guards not be removed? How was the work rule communicated? Was the work rule enforced?

(2) Impossibility. Compliance with the requirements of a standard is:

(a) Functionally impossible or would prevent performances of required work; and

(b) There are no alternative means of employee protection.

EXAMPLE: During the course of the inspection an unguarded table saw is observed. The employer states that the nature of its work makes a guard unworkable. Facts which the CSHO shall document may include: Would a guard make performance of the work impossible or merely more difficult? Could a guard be used part of the time? Has the employer attempted to use guards? Has the employer considered alternative means or methods of avoiding or reducing the hazard?

C. 8. c. (3) <u>Greater Hazard</u>. Compliance with a standard would result in greater hazards to employees than noncompliance and:

(a) There are no alternative means of employee protection; and

(b) An application of a variance would be inappropriate.

EXAMPLE: The employer indicates that a saw guard had been removed because it caused particles to be thrown into the operator's face. Facts which the CSHO shall consider may include: Was the guard used properly? Would a different type of guard eliminate the problem? How often was the operator struck by particles and what kind of injuries resulted? Would safety glasses, a face mask, or a transparent sheif attached to the saw prevent injury? Was operator technique at fault and did the employer attempt to correct it? Was a variance sought?

(4) <u>Multiemployer Worksites</u>. Refer to C.6. of this chapter.

CHAPTER IV

POST-INSPECTION PROCEDURES

A. Abatement.

1. Period. The abatement period shall be the shortest interval within which the employer can **reasonably** be expected to correct the violation. An abatement date shall be set forth in the citation as a specific date, not a number of days. When the abatement period is very short (i.e., 5 working days or less) and it is uncertain when the employer will receive the citation, the abatement date shall be set so as to allow for a mail delay and the agreed-upon abatement time. When abatement has been witnessed by the CSHO during the inspection, the abatement period shall be "Corrected During Inspection" on the citation.

2. Reasonable Abatement Date. The establishment of the shortest practicable abatement date requires the exercise of professional judgment on the part of the CSHO.

   NOTE: Abatement periods exceeding 30 calendar days should not normally be necessary, particularly for safety violations. Situations may arise, however, especially for health violations, where extensive structural changes are necessary or where new equipment or parts cannot be delivered within 30 calendar days. When an initial abatement date is granted that is in excess of 30 calendar days, the reason, if not self-evident, shall be documented in the case file.

3. Verification of Abatement. The Area Director is responsible for determining if abatement has been accomplished. When abatement is not accomplished during the inspection or the employer does not notify the Area Director by letter of the abatement, verification shall be determined by telephone and documented in the case file.

   NOTE: If the employer's abatement letter indicates that a condition has not been abated, but the date has passed, the Area Director shall contact the employer for an explanation. The Area Director shall explain Petition for Modification of Abatement (PMA) procedures to the employer, if applicable.

4. Effect of Contest Upon Abatement Period. In situations where an employer contests either (1) the period set for abatement or (2) the citation itself, the abatement period generally shall be considered not to have begun until there has been an affirmation of the citation and abatement period. In accordance with the Act, the abatement period begins when a final order of the Review Commission is issued, and this abatement period is not tolled while an appeal to the court is ongoing

unless the employer has been granted a stay. In situations where there is an employee contest of the abatement date, the abatement requirements of the citation remain unchanged.

A. 4.   a.   Where an employer has contested only the proposed penalty, the abatement period continues to run unaffected by the contest.

b.   Where the employer does not contest, he must abide by the date set forth in the citation even if such date is within the 15-working-day notice of contest period. Therefore, when the abatement period designated in the citation is 15 working days or less and a notice of contest has not been filed, a follow-up inspection of the worksite may be conducted for purposes of determining whether abatement has been achieved within the time period set forth in the citation. A failure to abate notice may be issued on the basis of the CSHO's findings.

c.   Where the employer has filed a notice of contest to the initial citation within the contest period, the abatement period does not begin to run until the entry of a final Review Commission order. Under these circumstances, any follow-up inspection within the contest period shall be discontinued and a failure to abate notice shall not be issued.

NOTE: There is one exception to the above rule. If an early abatement date has been designated in the initial citation and it is the opinion of the CSHO and/or the Area Director that a situation classified as imminent danger is presented by the cited condition, appropriate imminent danger proceedings may be initiated notwithstanding the filing of a notice of contest by the employer.

d.   If an employer contests an abatement date in good faith, a Failure to Abate Notice shall not be issued for the item contested until a final order affirming a date is entered, the new abatement period, if any, has been completed, and the employer has still failed to abate.

5.   Long-Term Abatement Date for Implementation of Feasible Engineering Controls. Long-term abatement is abatement which will be completed more than one year from the citation issuance date. In situations where it is difficult to set a specific abatement date when the citation is originally issued; e.g., because of extensive redesign requirements consequent upon the employer's decision to implement feasible engineering controls and uncertainty as to when the job can be finished, the CSHO shall discuss the problem with the employer at the closing conference and, in appropriate cases, shall encourage the employer to seek an informal conference with the Area Director.

a.   <u>Final Abatement Date</u>. The CSHO and the Assistant Area Director shall make their best judgment as to a reasonable abatement date. A specific date for final abatement shall, in all cases, be included in the citation. The employer shall not be permitted to propose an abatement plan setting its own abatement dates. If necessary, an appropriate petition may be submitted later by the employer to the Area Director to modify the abatement date. (See D.2. of this chapter for PMA's.)

b.   <u>Employer Abatement Plan</u>. The employer is required to submit an abatement plan outlining the anticipated long-term abatement procedures.

   NOTE: A statement agreeing to provide the affected Area Offices with written periodic progress reports shall be part of the long-term abatement plan.

A. 6.   <u>Feasible Administrative, Work Practice and Engineering Controls</u>. Where applicable, the CSHO shall discuss control methodology with the employer during the closing conference.

a. <u>Definitions</u>.

   (1) <u>Engineering Controls</u>. Engineering controls consist of substitution, isolation, ventilation and equipment modification.

   (2) <u>Administrative Controls</u>. Any procedure which significantly limits daily exposure by control or manipulation of the work schedule or manner in which work is performed is considered a means of administrative control. The use of personal protective equipment is <u>not</u> considered a means of administrative control.

   (3) <u>Work Practice Controls</u>. Work practice controls are a type of administrative controls by which the employer modifies the manner in which the employee performs assigned work. Such modification may result in a reduction of exposure through such methods as changing work habits, improving sanitation and hygiene practices, or making other changes in the way the employee performs the job.

   (4) <u>Feasibility</u>. Abatement measures required to correct a citation item are feasible when they can be accomplished by the employer. The CSHO, following current directions and guidelines, shall inform the employer, where appropriate, that a determination will be made as to whether engineering or administrative controls are feasible.

      (a)   <u>Technical Feasibility</u>. Technical feasibility is the existence of technical know-how as to materials and methods available or adaptable to specific circumstances which can be applied to cited violations with a reasonable possibility that employee exposure to occupational hazards will be reduced.

    (b) <u>Economic Feasibility</u>. Economic feasibility means that the employer is financially able to undertake the measures necessary to abate the citations received.

> **NOTE:**   If an employer's level of compliance lags significantly behind that of its industry, allegations of economic infeasibility will not be accepted.

A. 6. b.   <u>Responsibilities</u>.

    (1)   The CSHO shall document the underlying facts which give rise to an employer's claim of infeasibility.

        (a)   When economic infeasibility is claimed the CSHO shall inform the employer that, although the cost of corrective measures to be taken will generally not be considered as a factor in the issuance of a citation, it may be considered during an informal conference or during settlement negotiations.

        (b)   Serious issues of feasibility should be referred to the Area Director for determination.

    (2)   The Area Director is responsible for making determinations that engineering or administrative controls are or are not feasible.

    c.   <u>Reducing Employee Exposure</u>. Whenever feasible engineering, administrative or work practice controls can be instituted even though they are not sufficient to eliminate the hazard (or to reduce exposure to or below the permissible exposure limit (PEL)). Nonetheless, they are required in conjunction with personal protective equipment to reduce exposure to the lowest practical level.

B. <u>Citations</u>.

    1. <u>Issuing Citations</u>.

        a.   <u>Sending Citations to the Employer</u>. Citations shall be sent by certified mail; hand delivery of citations to the employer or an appropriate agent of the employer may be substituted for certified mailing if it is believed that this method would be more effective. A signed receipt shall be obtained when-ever possible; otherwise the circumstances of delivery shall be documented in the file.

b.   <u>Sending Citations to the Employee</u>. Citations shall be mailed to employee representatives no later than one day after the citation is sent to the employer. Citations shall also be mailed to any employee upon request.

c.   <u>Followup Inspections</u>. If a followup inspection reveals a failure to abate, the time specified for abatement has passed, and no notice of contest has been filed, a Notification of Failure to Abate Alleged Violation (OSHA-2B) may be issued immediately without regard to the contest period of the initial citation.

B. 2.   <u>Amending or Withdrawing Citation and Notification of Penalty in Part or In Its Entirety</u>.

a.   <u>Citation Revision Justified</u>. Amendments to or withdrawal of a citation shall be made when information is presented to the Area Director which indicates a need for such revision under certain conditions which may include:

   (1)   Administrative or technical error.

      (a)   Citation of an incorrect standard.

      (b)   Incorrect or incomplete description of the alleged violation.

   (2)   Additional facts establish a valid affirmative defense.

   (3)   Additional facts establish that there was no employee exposure to the hazard.

   (4)   Additional facts establish a need for modification of the correction date, or the penalty, or reclassification of citation items.

b.   <u>Citation Revision Not Justified</u>. Amendments to or withdrawal of a citation shall not be made by the Area Director under certain conditions which include:

   (1)   Valid notice of contest received.

   (2)   The 15 working days for filing a notice of contest has expired and the citation has become a final order.

(3) Employee representatives have not been given the opportunity to present their views unless the revision involves only an administrative or technical error.

(4) Editorial and/or stylistic modifications.

B. 2. c. <u>Procedures for Amending or Withdrawing Citations</u>. The following procedures are to be followed in amending or withdrawing citations. The instructions contained in this section, with appropriate modification, are also applicable to the amendment of the Notification of Failure to Abate Alleged Violation, OSHA-2B Form:

(1) Withdrawal of or modifications to the citation and notification of penalty, shall normally be accomplished by means of an informal settlement agreement (ISA). (See D.4.b. of this chapter for further information in ISA's).

(2) Changes initiated by the Area Director without an informal conference are exceptions. In such cases the procedures given below shall be followed:

(a) If proposed amendments to citation items change the classification of the items; e.g., serious to other-than-serious, the original citation items shall be withdrawn and new, appropriate citation items issued.

(b) The amended Citation and Notification of Penalty Form (OSHA-2) shall clearly indicate that:

<u>1</u> The employer is obligated under the Act to post the amendment to the citation along with the original citation until the amended violation has been corrected or for 3 working days, whichever is longer;

<u>2</u> The period of contest of the amended portions of the OSHA-2 will begin from the day following the date of receipt of the amended Citation and Notification of Penalty; and

<u>3</u> The contest period is not extended as to the unamended portions of the original citation.

(c) A copy of the original citation shall be attached to the amended Citation and Notification of Penalty Form when the amended form is forwarded to the employer.

B. 2. c.   (2)   (d)   When circumstances warrant it, a citation may be withdrawn in its entirety by the Area Director. Justifying documentation shall be placed in the case file. If a citation is to be withdrawn, the following procedures apply:

<u>1</u>   A letter withdrawing the Citation and Notification of Penalty shall be sent to the employer. The letter shall refer to the original citation and penalty, state that they are withdrawn and direct that the letter be posted by the employer for 3 working days in those locations where the original citation was posted.

<u>2</u>   When applicable to the specific situation (e.g., an employee representative participated in the walkaround inspection, the inspection was in response to a complaint signed by an employee or an employee representative, or the withdrawal resulted from an informal conference or settlement agreement in which an employee representative exercised the right to participate), a copy of the letter shall also be sent to the employee or the employee representative as appropriate.

C. <u>Penalties</u>.

1. <u>General Policy</u>. The penalty structure provided under Section 17 of the Act is designed primarily to provide an incentive toward correcting violations voluntarily, not only to the offending employer but, more especially, to other employers who may be guilty of the same infractions of the standards or regulations.

     a.   While penalties are not designed primarily as punishment for violations, the Congress has made clear its intent that penalty amounts should be sufficient to serve as an effective deterrent to violations.

     b.   Large proposed penalties, therefore, serve the public purpose intended under the Act; and criteria guiding approval of such penalties by the Assistant Secretary are based on meeting this public purpose. (See OSHA Instruction CPL 2.80.)

     c.   The penalty structure outlined in this section is designed as a general guideline. The Area Director may deviate from this guideline if warranted, to achieve the appropriate deterrent effect.

2. <u>Civil Penalties.</u>

     a.   <u>Statutory Authority</u>. Section 17 provides the Secretary with the statutory authority to propose civil penalties for violations of the Act.

C. 2. a. (1) Section 17(b) of the Act provides that any employer who has received a citation for an alleged violation of the Act which is determined to be of a serious nature shall be assessed a civil penalty of up to $7,000 for each violation. (See OSHA Instruction CPL 2.51H, or the most current version, for congressional exemptions and limitations placed on penalties by the Appropriations Act.)

(2) Section 17(c) provides that, when the violation is specifically determined not to be of a serious nature, a proposed civil penalty of up to $7,000 may be assessed for each violation.

(3) Section 17(i) provides that, when a violation of a posting requirement is cited, a civil penalty of up to $7,000 shall be assessed.

b. <u>Minimum Penalties</u>. The following guidelines apply:

(1) The proposed penalty for any willful violation shall not be less than $5,000 for other-than-serious and regulatory violations, and shall not be less than $25,000 for serious violations. The $5,000 penalty is a statutory minimum and not subject to administrative discretion.

(2) When the adjusted proposed penalty for an other-than-serious violation (citation item) would amount to less than $100, no penalty shall be proposed for that violation.

(3) When, however, there is a citation item for a posting violation, this minimum penalty amount does not apply with respect to that item since penalties for such items are mandatory under the Act.

(4) When the adjusted proposed penalty for a serious violation (citation item) would amount to less than $100, a $100 penalty shall be proposed for that violation.

c. <u>Penalty Factors</u>. Section 17(j) of the Act provides that penalties shall be assessed on the basis of four factors:

(1) The gravity of the violation,

(2) The size of the business,

(3) The good faith of the employer, and

(4) The employer's history of previous violations.

C. 2. d.   <u>Gravity of Violation</u>. The gravity of the violation is the primary consideration in determining penalty amounts. It shall be the basis for calculating the basic penalty for both serious and other violations. To determine the gravity of a violation the following two assessments shall be made:

(1)   The severity of the injury or illness which could result from the alleged violation.

(2)   The probability that an injury or illness could occur as a result of the alleged violation.

e.   <u>Severity Assessment</u>. The classification of the alleged violations as serious or other-than-serious, in accordance with the instructions in Chapter III, C.2., is based on the severity of the injury or illness that could result from the violation. This classification constitutes the first step in determining the gravity of the violation. A severity assessment shall be assigned to a hazard to be cited according to the most serious injury or illness which could reasonably be expected to result from an employee's exposure as follows:

(1)   <u>High Severity</u>: Death from injury or illness; injuries involving permanent disability; or chronic, irreversible illnesses.

(2)   <u>Medium Severity</u>: Injuries or temporary, reversible illnesses resulting in hospitalization or a variable but limited period of disability.

(3)   <u>Low Severity</u>: Injuries or temporary, reversible illnesses not resulting in hospitalization and requiring only minor supportive treatment.

(4)   <u>Minimal Severity</u>: Other-than-serious violations. Although such violations reflect conditions which have a direct and immediate relationship to the safety and health of employees, the injury or illness most likely to result would probably not cause death or serious physical harm.

f.   <u>Probability Assessment</u>. The probability that an injury or illness will result from a hazard has no role in determining the classification of a violation but does affect the amount of the penalty to be proposed.

(1)   <u>Categorization</u>. Probability shall be categorized either as greater or as lesser probability.

     (a)   Greater probability results when the likelihood that an injury or illness will occur is judged to be relatively high.

     (b)   Lesser probability results when the likelihood that an injury or illness will occur is judged to be relatively low.

C. 2. f. (2) Violations. The following circumstances may normally be considered, as appropriate, when violations likely to result in injury or illness are involved:

(a) Number of workers exposed.

(b) Frequency of exposure or duration of employee overexposure to contaminants.

(c) Employee proximity to the hazardous conditions.

(d) Use of appropriate personal protective equipment (PPE).

(e) Medical surveillance program.

(f) Other pertinent working conditions.

(3) Final Probability Assessment. All of the factors outlined above shall be considered together in arriving at a final probability assessment. When strict adherence to the probability assessment procedures would result in an unreasonably high or low gravity, the probability may be adjusted as appropriate based on professional judgment. Such decisions shall be adequately documented in the case file.

g. Gravity-Based Penalty. The gravity-based penalty (GBP) is an unadjusted penalty and is calculated in accordance with the following procedures:

(1) The GBP for each violation shall be determined based on an appropriate and balanced professional judgment combining the severity assessment and the final probability assessment.

(2) For serious violations, the GBP shall be assigned on the basis of the following scale:

Severity	Probability	GBP	Gravity
High	Greater	$5,000	high ($5,000+)
Medium	Greater	$3,500----	
Low	Greater	$2,500	|--moderate
High	Lesser	$2,500	|
Medium	Lesser	$2,000----	
Low	Lesser	$1,500	low

NOTE: The gravity of a violation is defined by the GBP.

- o A **high gravity** violation is one with a GBP of $5,000 or greater.

- o A **moderate gravity** violation is one with GBP of $2,000 to $3,500.

- o A **low gravity** violation is one with a GBP of $1,500.

C. 2. g. (3) The highest gravity classification (high severity and greater probability) shall normally be reserved for the most serious violative conditions such as those situations involving danger of death or extremely serious injury or illness. If the Area Director determines that it is appropriate to achieve the necessary deterrent effect, a GBP of $7,000 may be proposed. The reasons for this determination shall be documented in the case file.

(4) For other-than-serious safety and health violations, there is no severity assessment.

(5) The Area Director may authorize a penalty between $1,000 and $7,000 for an other-than-serious violation when it is determined to be appropriate to achieve the necessary deterrent effect. The reasons for such a determination shall be documented in the case file.

Probability	GBP
Greater	$1,000 - $7,000
Lesser	$0

(6) A GBP may be assigned in some cases without using the severity and the probability assessment procedures outlined in this section when these procedures cannot appropriately be used.

(7) The Penalty Table (Table IV-1) may be used for determining appropriate adjusted penalties for serious and other-than-serious violations.

h. Gravity Calculations for Combined or Grouped Violations. Combined or grouped violations will normally be considered as one violation and shall be assessed one GBP. The following procedures apply to the calculation of penalties for combined and grouped violations:

(1) The severity and the probability assessments for combined violations shall be based on the instance with the highest gravity. It is not neces-

sary to complete the penalty calculations for each instance or subitem of a combined or grouped violation if it is clear which instance will have the highest gravity.

C. 2. h.   (2)   For grouped violations, the following special guidelines shall be adhered to:

(a)   Severity Assessment. There are two considerations to be kept in mind in calculating the severity of grouped violations:

1   The severity assigned to the grouped violation shall be no less than the severity of the most serious reasonably predictable injury or illness that could result from the violation of any single item.

2   If a more serious injury or illness is reasonably predictable from the grouped items than from any single violation item, the more serious injury or illness shall serve as the basis for the calculation of the severity factor of the grouped violation.

(b)   Probability Assessment. There are two considerations to be kept in mind in calculating the probability of grouped violations:

1   The probability assigned to the grouped violation shall be no less than the probability of the item which is most likely to result in an injury or illness.

2   If the overall probability of injury or illness is greater with the grouped violation than with any single violation item, the greater probability of injury or illness shall serve as the basis for the calculation of the probability assessment of the grouped violation.

(3)   In egregious cases an additional factor of up to the number of violation instances may be applied. Such cases shall be handled in accordance with OSHA Instruction CPL 2.80. Penalties calculated with this additional factor shall not be proposed without the concurrence of the Assistant Secretary. (See also C.2.k.(2)(c)4 of this chapter.)

i.   Penalty Adjustment Factors. The GBP may be reduced by as much as 95 per cent depending upon the employer's "good faith," "size of business," and "history of previous violations." Up to 60-percent reduction is permitted for size; up to 25- percent reduction for good faith, and 10-percent for history.

C. 2. i.   (1)   Since these adjustment factors are based on the general character of a business and its safety and health performance, the factors generally shall be calculated only once for each employer. After the classification and probability ratings have been determined for each violation, the adjustment factors shall be applied subject to the limitations indicated in the following paragraphs.

  (2)   Penalties assessed for violations that are classified as high severity and greater probability shall be adjusted only for size and history.

  (3)   Penalties assessed for violations that are classified as repeated shall be adjusted only for size.

  (4)   Penalties assessed for regulatory violations, which are classified as willful, shall be adjusted for size. Penalties assessed for serious violations, which are classified as willful, shall be adjusted for size and history.

     NOTE: If one violation is classified as willful, no reduction for good faith can be applied to any of the violations found during the same inspection. The employer cannot be willfully in violation of the Act and at the same time, be acting in good faith.

  (5)   The rate of penalty reduction for size of business, employer's good faith and employer's history of previous violations shall be calculated on the basis of the criteria described in the following paragraphs:

    (a)   <u>Size</u>. A maximum penalty reduction of 60 percent is permitted for small businesses. "Size of business" shall be measured on the basis of the maximum number of employees of an employer at all workplaces at any one time during the previous 12 months.

      <u>1</u>   The rates of reduction to be applied are as follows:

Employees	Percent reduction
1-25	60
26-100	40
101-250	20
251 or more	None

      <u>2</u>   When a small business (1-25 employees) has one or more serious violations of high gravity or a number of serious violations of moderate gravity, indicating a lack of concern for

employee safety and health, the CSHO may recommend that only a partial reduction in penalty shall be permitted for size of business.

C. 2. i.   (5)   (b)   Good Faith.   A penalty reduction of up to 25 percent, based on the CSHO's professional judgment, is permitted in recognition of an employer's "good faith".

1   The 25% credit for "good faith" normally requires a written safety and health program. In exceptional cases, the compliance officer may recommend the full 25% for a smaller employer (1-25 employees) who has implemented an efficient safety and health program, but has not reduced it to writing.

a   Provides for appropriate management commitment and employee involvement; worksite analysis for the purpose of hazard identification; hazard prevention and control measures; and safety and health training.

NOTE: One example of a framework for such a program is given in OSHA's voluntary "Safety and Health Program Management Guidelines" (Federal Register, Vol. 54, No. 16, January 26, 1989, pp. 3904- 3916, or later revisions as published).

b   Has deficiencies that are only incidental.

2   A reduction of 15 percent shall normally be given if the employer has a documentable and effective safety and health program, but with more than only incidental deficiencies.

3   No reduction shall be given to an employer who has no safety and health program or where a willful violation is found.

4   Only these percentages (15% or 25%) may be used to reduce penalties due to the employer's good faith. No intermediate percentages shall be used.

(c)   History. A reduction of 10 percent shall be given to employers who have not been cited by OSHA for any serious, willful, or repeated violations in the past three years.

(d)   Total. The total reduction will normally be the sum of the reductions for each adjustment factors.

C. 2. j.   <u>Effect on Penalties If Employer Immediately Corrects or Initiates Corrective Action</u>. Appropriate penalties will be proposed with respect to an alleged violation even though, after being informed of such alleged violation by the CSHO, the employer immediately corrects or initiates steps to correct the hazard.

   k.   <u>Failure to Abate</u>. A Notification of Failure to Abate an Alleged Violation (OSHA-2B) shall be issued in cases where violations have not been corrected as required.

   (1)   <u>Failure to Abate</u>. Failure to abate penalties shall be applied when an employer has not corrected a previously cited violation which had become a final order of the Commission. Citation items become final order of the Review Commission when the abatement date for that item passes, if the employer has not filed a notice of contest prior to that abatement date. See D.5. of this chapter for guidance on determining final dates of settlements and Review Commission orders.

   (2)   <u>Calculation of Additional Penalties</u>. A GBP for unabated violations is to be calculated for failure to abate a serious or other-than-serious violation on the basis of the facts noted upon reinspection. This recalculated GBP, however, shall not be less than that proposed for the item when originally cited, except as provided in C.2.k.(4), below.

   (a)   In those instances where no penalty was initially proposed, an appropriate penalty shall be determined after consulting with the Assistant Area Director. In no case shall the unadjusted penalty be less than $1,000 per day.

   (b)   Only the adjustment factor for size--based upon the circumstances noted during the reinspection--shall then be applied to arrive at the daily proposed penalty.

   (c)   The daily proposed penalty shall be multiplied by the number of calendar days that the violation has continued unabated, except as provided below:

   <u>1</u>   The number of days unabated shall be counted from the day following the abatement date specified in the citation or the final order. It will include all calendar days between that date and the date of reinspection, excluding the date of reinspection.

C. 2. k. (2) (c) <u>2</u>   Normally the maximum total proposed penalty for failure to abate a particular violation shall not exceed 30 times the amount of the daily proposed penalty.

<u>3</u>   At the discretion of the Area Director, a lesser penalty may be proposed with the reasons for doing so (e.g., achievement of an appropriate deterrent effect) documented in the case file.

<u>4</u>   If a penalty in excess of the normal maximum amount of 30 times the amount of the daily proposed penalty is deemed appropriate by the Area Director, the case shall be treated under the violation-by-violation (egregious) penalty procedures established in OSHA Instruction CPL 2.80.

(3) Partial Abatement.

(a) When the citation has been partially abated, the Area Director may authorize a reduction of 25 percent to 75 percent to the amount of the proposed penalty calculated as outlined in C.2.k.(2), above.

(b) When a violation consists of a number of instances and the follow-up inspection reveals that only some instances of the violation have been corrected, the additional daily proposed penalty shall take into consideration the extent that the violation has been abated.

EXAMPLE: Where 3 out of 5 instances have been corrected, the daily proposed penalty (calculated as outlined in C.2.k.(2), above, without regard to any partial abatement) may be reduced by 60 per cent.

(4) Good Faith Effort to Abate. When the CSHO believes, and so documents in the case file, that the employer has made a good faith effort to correct the violation and had good reason to believe that it was fully abated, the Area Director may reduce or eliminate the daily proposed penalty that would otherwise be justified.

l. Repeated Violations. Section 17(a) of the Act provides that an employer who **repeatedly** violates the Act may be assessed a civil penalty of not more than $70,000 for each violation.

(1) Gravity-Based Penalty Factors. Each violation shall be classified as serious or other-than-serious. A GBP shall then be calculated for repeated violations based on facts noted during the current inspection. Only the adjustment factor for size, appropriate to the facts at the time of the reinspection, shall be applied.

C. 2. l.   (2) <u>Penalty Increase Factors</u>. The amount of the increased penalty to be assessed for a repeated violation shall be determined by the size of the employer.

    (a) <u>Smaller Employers</u>. For employers with 250 or fewer employees, the GBP shall be doubled for the first repeated violation and quintupled if the violation has been cited twice before. If the Area Director determines that it is appropriate to achieve the necessary deterrent effect, the GBP may be multiplied by **10**.

    (b) <u>Larger Employers</u>. For employers with more than 250 employees, the GBP shall be multiplied by **5** for the first repeated violation and multiplied by **10** for the second repeated violation.

   (3) <u>Other-Than-Serious, No Initial Penalty</u>. For a repeated other-than-serious violation that otherwise would have no initial penalty, a GBP penalty of $200 shall be assessed for the first repeated violation, $500 if the violation has been cited twice before, and $1,000 for a third repetition.

    NOTE: This penalty will not have the penalty increase factors applied as discussed under C.2.l.(2).

   (4) <u>Regulatory Violations</u>. For repeated instances of regulatory violations, the initial penalty shall be doubled for the first repeated violation and quintupled if the violation has been cited twice before. If the Area Director determines that it is appropriate to achieve the necessary deterrent effect, the initial penalty may be multiplied by **10**.

    NOTE: See Chapter III, C.2.f., for additional guidance on citing repeated violations.

  m. <u>Willful Violations</u>. Section 17(a) of the Act provides that an employer who willfully violates the Act may be assessed a civil penalty of not more than $70,000 but not less than $5,000 for each violation.

   (1) <u>Gravity-Based Penalty Factors</u>. Each willful violation shall be classified as serious or other-than-serious.

C. 2. m. (1) (a) <u>Serious Violations</u>. For willful serious violations, a gravity of **high**, **medium**, or **low** shall be assigned based on the GBP of the underlying serious violation, as described at C.2.g.(2).

    <u>1</u>   The adjustment factor for size shall be applied at **one-half** of the values stated at C.2.i.(5)(a)<u>1</u>; i.e., a reduction of 30 percent (1-25 employees), 20 percent (26-100 employees), 10 percent (101-250 employees), or no reduction (251 or more employees).

    <u>2</u>   The adjustment factor for history shall be applied as described at C.2.i.(5)(c); i.e., a reduction of 10 percent shall be given to employers who have not been cited by OSHA for any serious, willful, or repeated violations in the past 3 years. There shall be no adjustment for good faith.

    <u>3</u>   The proposed penalty shall then be determined from the table below:

	Penalties to be proposed				
Total percentage reduction for size and/or history	0%	10%	20%	30%	40%
High Gravity	$70,000	$63,000	$56,000	$49,000	$42,000
Moderate Gravity	$55,000	$49,500	$44,000	$38,500	$33,000
Low Gravity	$40,000	$36,000	$32,000	$28,000	$25,000[1]

[1] See C.2.m.(1)(a)<u>1</u> below.

    <u>4</u>   In no case shall the proposed penalty be less than $25,000.

    (b)   <u>Other-Than-Serious Violations</u>. For willful other-than-serious violations, the minimum willful penalty of $5,000 shall be assessed.

C. 2. m.   (2) <u>Regulatory Violations</u>. In the case of regulatory violations (see C.2.n., below) that are determined to be willful, the unadjusted initial penalty shall be multiplied by **10**. In no event shall the penalty, after adjustment for size, be less than $5,000.

n.   <u>Violation of 29 CFR Parts 1903 and 1904 Regulatory Requirements</u>. Except as provided in the Appropriations Act, Section 17 of the Act provides that an employer who violates any of the posting requirements shall be assessed a civil penalty of up to $7,000 for each violation and may be assessed a like penalty for recordkeeping violations.

   (1) <u>General Application</u>. Unadjusted penalties for regulatory violations, including posting requirements, shall have the adjustment factors for size and history applied (excluding willful violations, see C.2.m.(2), above).

   (2) <u>Posting Requirements</u>. Penalties for violation of posting requirements shall be proposed as follows:

      (a) <u>OSHA Notice (Poster)</u>. If the employer has not displayed (posted) the notice furnished by the Occupational Safety and Health Administration as prescribed in 29 CFR §1903.2 (a), an other-than-serious citation shall normally be issued. The unadjusted penalty for this alleged violation shall be $1,000 provided that the employer has received a copy of the poster or had knowledge of the requirement.

      (b) <u>Annual Summary</u>. If an employer fails to post the summary portion of the OSHA-200 Form during the month of February as required by 29 CFR §1904.5(d)(1), and/or fails to complete the summary prior to February 1, as required by 29 CFR §1904.5(b), even if there have been no injuries, an other-than-serious citation shall be issued. The unadjusted penalty for this violation shall be $1,000.

      (c) <u>Citation</u>. If an employer received a citation that has not been posted as prescribed in 29 CFR §1903.16, an other-than-serious citation shall normally be issued. The unadjusted penalty shall be $3,000.

C. 2. n. (3) <u>Reporting and Recordkeeping Requirements</u>. Section 17(c) of the Act provides that violations of the recordkeeping and reporting requirements may be assessed civil penalties of up to $7,000 for each violation.

    (a) <u>OSHA-200 Form</u>. If the employer does not maintain the Log and Summary of Occupational Injuries and Illnesses, OSHA-200 Form, as prescribed in 29 CFR Part 1904, an other-than-serious citation shall be issued. There shall be an unadjusted penalty of $1,000 for each year the form was not maintained, for each of the preceding 3 years.

        <u>1</u>   When no recordable injuries or illnesses have occurred at a workplace during the current calendar year, the OSHA 200 need not be completed until the end of the calendar year for certification of the summary.

        <u>2</u>   An OSHA-200 with significant deficiencies shall be considered as not maintained.

    (b) <u>OSHA-101 Forms</u>. If the employer does not maintain the Supplementary Record, OSHA 101 Form (or equivalent), as prescribed in 29 CFR Part 1904, an other-than-serious citation shall be issued. There shall be an unadjusted penalty of $1000 for each OSHA-101 Form not maintained.

        <u>1</u>   A penalty of $1000 for each OSHA-101 Form not maintained at all up to a maximum of $7000.

        <u>2</u>   A penalty of $1,000 for each OSHA-101 Form inaccurately maintained up to a maximum of $3000.

        <u>3</u>   Minor inaccuracies shall be cited, but with no penalties.

        <u>4</u>   If large numbers of violations or other circumstances indicate that the violations are willful, then other penalties including, violation-by-violation, may be applied.

C. 2. n. (3) (c) <u>Reporting</u>. Employers are required to report either orally or in writing to the nearest Area Office within 8 hours, any occurrence of an employment accident which is fatal to one or more employees or which results in the hospitalization of three or more employees.

<u>1</u> An other-than-serious citation shall be issued for failure to report such an occurrence. The unadjusted penalty shall be $5,000.

<u>2</u> If the Area Director determines that it is appropriate to achieve the necessary deterrent effect, an unadjusted penalty of $7,000 may be assessed.

<u>3</u> If the Area Director becomes aware of an incident required to be reported under 29 CFR §1904.8 through some means other than an employer report, prior to the elapse of the 8-hour reporting period and an inspection of the incident is made, a citable violation for failure to report does not exist.

(4) <u>Grouping</u>. Violations of the posting and record-keeping requirements which involve the same document (e.g., summary portion of the OSHA-200 Form was neither posted nor maintained) shall be grouped as an other-than- serious violation for penalty purposes. The unadjusted penalty for the grouped violations would then take on the highest dollar value of the individual items, which will normally be $1,000.

(5) <u>Access to Records</u>.

(a) <u>29 CFR Part 1904</u>. If the employer fails upon request to provide records required in §1904.2 for inspection and copying by any employee, former employee, or authorized representative of employees, a citation for violation of 29 CFR §1904.7(b)(1) shall normally be issued. The unadjusted penalty shall be $1,000 for each form not made available.

<u>1</u> Thus, if the OSHA-200 for the 3 preceding years is not made available, the unadjusted penalty would be $3,000.

<u>2</u> If the employer is to be cited for failure to maintain these records, no citation of §1904.7 shall be issued.

C. 2. n. (5) (b) <u>29 CFR §1910.20</u>. If the employer is cited for failing to provide records as required under 29 CFR §1910.20 for inspection and copying by any employee, former employee, or authorized representative of employees, an unadjusted penalty of $1,000 shall be proposed for each record; i.e., either medical record or exposure record, on an individual employee basis. A maximum $7,000 may be assessed for such violations. This policy does not preclude the use of violation-by-violation penalties where appropriate. (See OSHA Instruction CPL 2.80.)

EXAMPLE: If all the necessary evidence is established where an authorized employee representative requested exposure and medical records for 3 employees and the request was denied by the employer, a citation would be issued for 6 instances of violation of 29 CFR §1910.20, with an unadjusted penalty of $6,000.

(6) <u>Notification Requirements</u>. When an employer has received advance notice of an inspection and fails to notify the authorized employee representative as required by 29 CFR §1903.6, an other-than-serious citation shall be issued. The violation shall have an unadjusted penalty of $2,000.

OSHA Instruction CPL 2.103
SEP 2 6 1994
Office of General Industry Compliance Assistance

TABLE IV-1

PENALTY TABLE

Percent Reduction	PENALTY (in dollars)							
0	1,000	1,500	2,000	2,500	3,000	3,500	5,000	7,000
10	900	1,350	1,800	2,250	2,700	3,150	4,500	6,300
15	850	1,275	1,700	2,125	2,550	2,975	4,250*	5,950*
20	800	1,200	1,600	2,000	2,400	2,800	4,000	5,600
25	750	1,125	1,500	1,875	2,250	2,625	3,750*	5,250*
30	700	1,050	1,400	1,750	2,100	2,450	3,500	4,900
35	650	975	1,300	1,625	1,950	2,275	3,250*	4,550*
40	600	900	1,200	1,500	1,800	2,100	3,000	4,200
45	550	825	1,100	1,375	1,650	1,925	2,750*	3,850*
50	500	750	1,000	1,250	1,500	1,750	2,500	3,500
55	450	675	900	1,125	1,350	1,575	2,250*	3,150*
60	400	600	800	1,000	1,200	1,400	2,000	2,800
65	350	525	700	875	1,050	1,225	1,750*	2,450*
70	300	450	600	750	900	1,050	1,500	2,100
75	250	375	500	625	750	875	1,250*	1,750*
85	150	225	300	375	450	525	750*	1,050*
95	100**	100**	100	125	150	175	250*	350*

*   Starred figures represent penalty amounts that would not normally be proposed for high gravity serious violations because no adjustment for good faith is made in such cases. They may occasionally be applicable for other-than-serious violations where the Area Director has determined a high unadjusted penalty amount to be warranted.

** Administratively, OSHA will not issue a penalty less than $100 for a serious violation.

C. 3. <u>Criminal Penalties.</u>

    a.   The Act and the U.S. Code provide for criminal penalties in the following cases:

        (1)   Willful violation of an OSHA standard, rule, or order causing the death of an employee (Section 17(e)).

        (2)   Giving unauthorized advance notice. (Section 17(f).)

        (3)   Giving false information. (Section 17(g).)

        (4)   Killing, assaulting or hampering the work of a CSHO. (Section 17(h)(2).)

    b.   Criminal penalties are imposed by the courts after trials and not by the Occupational Safety and Health Administration or the Occupational Safety and Health Review Commission.

D. <u>Post-Citation Processes.</u>

    1.   <u>Informal Conferences.</u>

        a.   <u>General.</u> Pursuant to 29 CFR §1903.19, the employer, any affected employee or the employee representative may request an informal conference. When an informal conference is conducted, it shall be conducted within the 15 working day contest period. If the employer's intent to contest is not clear, the Area Director shall contact the employer for clarification.

        b.   <u>Procedures.</u> Whenever an informal conference is requested by the employer, an affected employee or the employee representative, both parties shall be afforded the opportunity to participate fully. If either party chooses not to participate in the informal conference, a reasonable attempt shall be made to contact that party to solicit their input prior to signing an informal settlement agreement if the adjustments involves more than the penalty. If the requesting party objects to the attendance of the other party, separate informal conferences may be held. During the conduct of a joint informal conference, separate or private discussions shall be permitted if either party so requests. Informal conferences may be held by any means practical.

            (1)   The employer shall be requested to complete and post the form found at the end of the informal conference letter until after the informal conference has been held.

(2)   Documentation of the Area Director's actions notifying the parties of the informal conference shall be placed in the case file.

D. 1. c.   <u>Participation by OSHA Officials</u>. The inspecting CSHOs and their Assistant Area Directors shall be notified of an upcoming informal conference and, if practicable, given the opportunity to participate in the informal conference (unless, in the case of the CSHO, the Area Director anticipates that only a penalty adjustment will result).

(1)   At the discretion of the Area Director, one or more additional OSHA employees (in addition to the Area Director) may be present at the informal conference. In cases in which proposed penalties total $100,000 or more, a second OSHA staff member shall attend the informal conference.

(2)   The Area Director shall ensure that notes are made indicating the basis for any decisions taken at or as a result of the informal conference. It is appropriate to tape record the informal conference and to use the tape recording in lieu of written notes.

d.   <u>Conduct of the Informal Conference</u>. The Area Director shall conduct the informal conference in accordance with the following guidelines:

(1)   <u>Opening Remarks</u>. The opening remarks shall include discussions of the following:

(a)   Purpose of the informal conference.

(b)   Rights of participants.

(c)   Contest rights and time restraints.

(d)   Limitations, if any.

(e)   Settlements of cases.

(f)   Other relevant information.

(g)   If the Area Director states any views on the legal merits of the employer's contentions, it should be made clear that those views are personal opinions only.

(2) <u>Closing</u>. At the conclusion of the discussion the main issues and potential courses of action shall be summarized. A copy of the summary, together with any other relevant notes or tapes of the discussion made by the Area Director, shall be placed in the case file.

D. 1. e. <u>Decisions</u>. At the end of the informal conference, the Area Director shall make a decision as to what action is appropriate in the light of facts brought up during the conference.

(1) Changes to citations, penalties or abatement dates normally shall be made by means of an informal settlement agreement in accordance with current OSHA procedures; the reasons for such changes shall be documented in the case file. For more detail on settlement agreements, see D.4.b., below.

(2) Employers shall be informed that they are required by 29 CFR §1903.19 to post copies of all amendments to the citation resulting from informal conferences. Employee representatives must also be provided with copies of such documents. This regulation covers amended citations, citation withdrawals and settlement agreements.

f. <u>Failure to Abate</u>. If the informal conference involves an alleged failure to abate, the Area Director shall set a new abatement date in the informal settlement agreement, documenting for the case file the time that has passed since the original citation, the steps that the employer has taken to inform the exposed employees of their risk and to protect them from the hazard, and the measures that will have to be taken to correct the condition.

2. <u>Petitions for Modification of Abatement Date (PMA)</u>. Title 29 CFR §1903.14a governs the disposition of PMAs. If the employer requests additional abatement time after the 15-working-day contest period has passed, the following procedures for PMAs are to be observed:

a. <u>Filing Date</u>. A PMA must be filed in writing with the Area Director who issued the citation no later than the close of the next working day following the date on which abatement was originally required.

(1) If a PMA is submitted orally, the employer shall be informed that OSHA cannot accept an oral PMA and that a written petition must be mailed by the end of the next working day after the abatement date. If there is not sufficient time to file a written petition, the employer shall be informed of the requirements below for late filing of the petition.

(2) A late petition may be accepted only if accompanied by the employer's statement of exceptional circumstances explaining the delay.

D. 2. b. Failure to Meet All Requirements. If the employer's letter does not meet all the requirements of §1903.14a(b)(1)-(5), the employer shall be contacted within 10 working days and notified of the missing elements. A reasonable amount of time for the employer to respond shall be specified during this contact with the employer.

(1) If no response is received or if the information returned is still insufficient, a second attempt (by telephone or in writing) shall be made. The employer shall be informed of the consequences of a failure to respond adequately; namely, that the PMA will not be granted and the employer may, consequently, be found in failure to abate.

(2) If the employer responds satisfactorily by telephone and the Area Director determines that the requirements for a PMA have been met, appropriate documentation shall be placed in the case file.

c. Delayed Decisions. Although OSHA policy is to handle PMAs as expeditiously as possible, there are cases where the Area Director's decision on the PMA is delayed because of deficiencies in the PMA itself, a decision to conduct a monitoring inspection and/or the need for Regional Office or National Office involvement. Requests for additional time (e.g., 45 days) for the Area Director to formulate a position shall be sent to the Review Commission through the Regional Solicitor. A letter conveying this request shall be sent at the same time to the employer and the employee representatives.

d. Area Office Position on the PMA. After 15 working days following the PMA posting, the Area Director shall determine the Area Office position, agreeing with or objecting to the request. This shall be done within 10 working days following the 15 working days (if additional time has not been requested from the Review Commission; in the absence of a timely objection, the PMA is automatically granted even if not explicitly approved). The following action shall be taken:

(1) If the PMA requests an abatement date which is two years or less from the issuance date of the citation, the Area Director has the authority to approve or object to the petition.

(2) Any PMA requesting an abatement date which is more than two years from the issuance date of the citation requires the approval of the Regional Administrator as well as the Area Director.

(3) If the PMA is approved, the Area Director shall notify the employer and the employee representatives by letter.

D. 2. d. (4) If supporting evidence justifies it (e.g., employer has taken no meaning-ful abatement action at all or has otherwise exhibited bad faith), the Area Director or the Regional Administrator, as appropriate and after consultation with the Regional Solicitor, shall object to the PMA. In such a case, all relevant documentation shall be sent to the Review Commission in accordance with 29 CFR §1903.14a(d). Both the employer and the employee representatives shall be notified of this action by letter, with return receipt requested.

   (a) The letters of notification of the objection shall be mailed on the same date that the agency objection to the PMA is sent to the Review Commission.

   (b) When appropriate, after consultation with the Regional Solicitor, a failure to abate notification may be issued in conjunction with the objection to the PMA.

e. Employee Objections. Affected employees or their representatives may file an objection in writing to an employer's PMA with the Area Director within 10 working days of the date of posting of the PMA by the employer or its service upon an authorized employee representative.

   (1) Failure to file such a written objection with the 10-working-day period constitutes a waiver of any further right to object to the PMA.

   (2) If an employee or an employee representative objects to the extension of the abatement date, all relevant documentation shall be sent to the Review Commission.

      (a) Confirmation of this action shall be mailed (return receipt requested) to the objecting party as soon as it is accomplished.

      (b) Notification of the employee objection shall be mailed (return receipt requested) to the employer on the same day that the case file is forwarded to the Commission.

3. Services Available to Employers. Employers requesting abatement assistance shall be informed that OSHA is willing to work with them even after citations have been issued.

4. Settlement of Cases By Area Directors.

   a. General. Area Directors are granted settlement authority, using the following policy guidelines to negotiate settlement agreements.

D. 4. a.   (1)   Except for egregious cases, or cases which affect other jurisdictions, Area Directors are authorized to enter into Informal Settlement Agreements with an employer before the employer files a written notice of contest.

NOTE: After the employer has filed a written notice of contest, the Area Director may proceed toward a Formal Settlement Agreement with the concurrence of the Regional Solicitor in cases where a settlement appears probable without the need for active participation by an attorney.

(2)   Area Directors are authorized to change abatement dates, to reclassify violations (e.g., willful to serious, serious to other-than-serious), and to modify or withdraw a penalty, a citation or a citation item if the employer presents evidence during the informal conference which convinces the Area Director that the changes are justified.

(a)   If an employer, having been cited as willfully or repeatedly violating the Act, decides to correct all violations, but wishes to purge himself or herself of the adverse public perception attached to a willful or repeated violation classification and is willing to pay all or almost all of the penalty and is willing to make significant additional concessions, then a Section 17 designation may be applicable. Decisions to make a Section 17 designation shall be based on whether the employer is willing to make significant concessions.

NOTE: Significant concessions may include the company entering into a corporate-wide settlement agreement subject to OSHA Instruction CPL 2.90, providing employee training of a specified type and frequency, hiring a qualified safety and health consultant and implementing the recommendations, effecting a comprehensive safety and health program, reporting new construction jobs or other worksites to OSHA, or waiving warrants for specified inspections/periods.

(b)   A Section 17 designation also may be considered if the employer has advanced substantial reasons why the original classification is questionable but is willing to pay the penalty as proposed.

NOTE: Where the original classification clearly was excessive, Section 17 is not appropriate. Instead, the citation shall be amended to the appropriate classification.

D. 4. **a.** (3) The Area Director has authority to actively negotiate the amount of penalty reduction, depending on the circumstances of the case and what improvements in employee safety and health can be obtained in return.

(4) Employers shall be informed that they are required by 29 CFR §1903.19 to post copies of all amendments or changes resulting from informal conferences. Employee representatives must also be provided with copies of such documents. This regulation covers amended citations, citation withdrawals and settlement agreements.

b. Pre-Contest Settlement (Informal Settlement Agreement). Pre-contest settlements generally will occur during, or immediately following, the informal conference and prior to the completion of the 15 working day contest period.

(1) If a settlement is reached during the informal conference, an Informal Settlement Agreement shall be prepared and the employer representative shall be invited to sign it. The Informal Settlement Agreement shall be effective upon signature by both the Area Director and the employer representative so long as the contest period has not expired. Both shall date the document as of the day of actual signature.

(a) If the employer representative requests more time to consider the agreement and if there is sufficient time remaining of the 15-working-day period, the Area Director shall sign and date the agreement and provide the signed original for the employer to study while considering whether to sign it. A letter explaining the conditions under which the agreement will become effective shall be given (or mailed by certified mail, return receipt requested) to the employer and a record kept in the case file.

(b) The Area Director shall sign and date the agreement and provide the original (in person, or by certified mail, return receipt requested) to the employer if any other circumstances warrant such action; the agreement may also be sent to the employer for signature, and returned to the Area Director, via facsimile if circumstances warrant.

<u>1</u> If the signed agreement is provided to the employer, a copy shall be kept in the case file and the employer informed in writing that no changes are to be made to the original by the employer without explicit prior authorization for such changes from the Area Director.

D. 4. b. (1) (b) <u>2</u>   In every case the Area Director shall give formal notice in writing to the employer that the citation will become final and unreviewable at the end of the contest period unless the employer either signs the agreement or files a written notice of contest.

<u>3</u>   If the employer representative wishes to make any changes to the text of the agreement, the Area Director must agree to and authorize the proposed changes prior to the expiration of the contest period.

<u>a</u>   If the changes proposed by the employer are acceptable to the Area Director, they shall be authorized and the exact language to be written into the agreement shall be worked out mutually. The employer shall be instructed to incorporate the agreed-upon language into the agreement, sign it and return it to the Area Office as soon as practicable by telefacsimile, if possible.

<u>b</u>   Annotations incorporating the exact language of any changes authorized by the Area Director shall be made to the retained copy of the agreement, and a dated record of the authorization shall be signed by the Area Director and placed in the case file.

<u>4</u>   Upon receipt of the Informal Settlement Agreement signed by the employer, the Area Director shall ensure that any modified text of the agreement is in accordance with the notations made in the case file.

<u>a</u>   If so, the citation record shall be updated in IMIS in accordance with current procedures.

<u>b</u>   If not, and if the variations substantially change the terms of the agreement, the agreement signed by the employer shall be considered as a notice of intent to contest and handled accordingly. The employer shall be so informed as soon as possible.

<u>5</u>   A reasonable time shall be allowed for return of the agreement from the employer.

<u>a</u>   After that time, if the agreement has still not been received, the Area Director shall presume that the employer is not going to sign the agreement; and the

citation shall be treated as a final order until such time as the agreement is received, properly signed prior to the expiration of the contest period.

b   The employer shall be required to certify that the informal settlement agreement was signed prior to the expiration of the contest period.

D. 4. b.   (2)   If the Area Director's settlement efforts are unsuccessful and the employer contests the citation, the Area Director shall state the terms of the final settlement offer in the case file.

c.   Procedures for Preparing the Informal Settlement Agreement. The Informal Settlement Agreement shall be prepared and processed in accordance with current OSHA policies and practices. For guidance for determining final dates of settlements and Review Commission orders see D.5., below.

d.   Post-Contest Settlement (Formal Settlement Agreement). Post-contest settlements will generally occur before the complaint is filed with the Review Commission.

(1)   Following the filing of a notice of contest, the Area Director shall, unless other procedures have been agreed upon, notify the Regional Solicitor when it appears that negotiations with the employer may produce a settlement. This shall normally be done at the time when the notice of contest transmittal memorandum is sent to the Regional Solicitor.

(2)   If a settlement is later requested by the employer with the Area Director, the Area Director shall communicate the terms of the settlement to the Regional Solicitor who will then draft the settlement agreement.

e.   Corporate-Wide Settlement Agreements. Corporate-wide Settlement Agreements (CSAs) may be entered into under special circumstances to obtain formal recognition by the employer of cited hazards and formal acceptance of the obligation to seek out and abate those hazards throughout all workplaces under its control. Guidelines, policies and procedures for entering into CSA negotiations are found in OSHA Instruction CPL 2.90.

5.   Guidance for Determining Final Dates of Settlements and Review Commission Orders.

a.   Citation/Notice of Penalty Not Contested. The Citation/Notice of Penalty and abatement date becomes a final order of the Commission on the date the 15-working-day contest period expires.

D. 5. b. <u>Citation/Notice of Penalty Resolved by Informal Settlement Agreement (ISA)</u>. The ISA becomes final, with penalties due and payable, 15 working days after the date of the last signature.

    NOTE: A later due date for payment of penalties may be set by the terms of the ISA.

    NOTE: The Review Commission does NOT review the ISA.

  c. <u>Citation/Notice of Penalty Resolved by Formal Settlement Agreement (FSA)</u>. The Citation/Notice of Penalty becomes final 30 days after docketing of the Administrative Law Judge's (ALJ's) <u>Order</u> "approving" the parties' stipulation and settlement agreement, assuming there is no direction for review. The Commission's <u>Notice of Docketing</u> specifies the date upon which the decision becomes a final order. If the FSA is "approved" by a Commission's <u>Order</u>, it will become final after 60 days.

    NOTE: A later due date for payment of penalties may be set by the terms of the FSA.

    NOTE: Settlement is permitted and encouraged by the Commission at any stage of the proceedings. (See 29 CFR §2200.100(a).)

  d. <u>Citation/Notice of Penalty Resolved by an ALJ Decision</u>. The ALJ decision/report becomes a final order of the Commission 30 days after **docketing** unless the Commission directs a review of the case. The Commission's <u>Notice of Docketing</u> specifies the date upon which the decision becomes a final order.

  e. <u>ALJ Decision is Reviewed by Commission</u>. According to Section 11 of the OSH Act, the Commission decision becomes final 60 days after the <u>Notice of Commission Decision</u> if no appeal has been filed with the U.S. Court of Appeals. The <u>Notice of Commission Decision</u> specifies the date the Commission decision is issued.

  f. <u>Commission Decision Reviewed by the U.S. Court of Appeals</u>. The U.S. Court of Appeals' decision becomes final 90 days after the entry of the judgment, if no appeal has been filed with the U.S. Supreme Court.

E. <u>Review Commission</u>.

  1. <u>Transmittal of Notice of Contest and Other Documents to Commission</u>.

    a. <u>Notice of Contest</u>. In accordance with the Occupational Safety and Health Review Commission (OSHRC) revised Rules of Procedure (51 F.R. 32020,

No. 173, September 8, 1986), the original notice of contest, together with copies of all relevant documents (all contested Citations and Notifications of Penalty and Notifications of Failure to Abate Alleged Violation, and proposed additional penalty) shall be transmitted by the Area Director to the OSHRC postmarked prior to the expiration of 15 working days after receipt of the notice of contest (29 CFR §2200.33). The Regional Solicitor shall be consulted in questionable cases.

E. 1. a.   (1)   The envelope that contained the notice of contest shall be retained in the case file with the postmark intact.

        (2)   Where the Area Director is certain that the notice of contest was not mailed; i.e., postmarked, within the 15-working-day period allowed for contest, the notice of contest shall be returned to the employer who shall be advised of the statutory time limitation. The employer shall be informed that OSHRC has no jurisdiction to hear the case because the notice of contest was not filed within the 15 working days allowed and, therefore, that the notice of contest will not be forwarded to the OSHRC. A copy of all untimely notices of contest shall be retained in the case file.

        (3)   If the notice of contest is submitted to the Area Director after the 15-working-day period, but the notice contests only the reasonableness of the abatement period, it shall be treated as a Petition for Modification of Abatement and handled in accordance with the instructions in D.2. of this chapter.

        (4)   If written communication is received from an employer containing objection, criticism or other adverse comment as to a citation or proposed penalty, which does not clearly appear to be a notice of contest, the Area Director shall contact the employer as soon as possible to clarify the intent of the communication. Such clarification must be obtained within 15 working days after receipt of the communication so that if, in fact, it is a notice of contest, the file may be forwarded to the Review Commission within the allowed time. The Area Director shall make a memorandum for the case file regarding the substance of this communication.

        (5)   If the Area Director determines that the employer intends the document to be a notice of contest, it shall be transmitted to the OSHRC in accordance with E.1.a., above. If the employer did not intend the document to be a notice of contest, it shall be retained in the case file with the memorandum of the contact with the employer. If no contact can be made with the employer, communications of the kind referred to in E.1.a.(4), above, shall be timely transmitted to the OSHRC.

(6) If the Area Director's contact with the employer reveals a desire for an informal conference, the employer shall be informed that an informal conference does not stay the running of the 15-working-day period for contest.

E. 1. b. <u>Documents to Executive Secretary</u>. The following documents are to be transmitted within the 15-working day time limit to the Executive Secretary, Occupational Safety and Health Review Commission, 1825 K Street, N.W., Washington, D.C. 20006:

> **NOTE:** In order to give the Regional Solicitor the maximum amount of time to prepare the information needed in filing a complaint with the Review Commission, the notice of contest and other documents shall not be forwarded to the Review Commission until the final day of the 15- working-day period.

(1) <u>All Notices of Contest</u>. The originals are to be transmitted to the Commission and a copy of each retained in the case file.

(2) <u>All Contested Citations and Notices of Proposed Penalty or Notice of Failure to Abate Issued in the Case</u>. The signed copy of each of these documents shall be taken from the case file and sent to the Commission after a copy of each is made and placed in the case file.

(3) <u>Certification Form</u>. The certification form shall be used for all contested cases and a copy retained in the case file. It is essential that the original of the certification form, properly executed, be transmitted to the Commission.

  (a) When listing the Region number in the heading of the form, do not use Roman numerals. Use 1, 2, 3, 4, 5, 6, 7, 8, 9, or 10. Insert "C" in the CSHO Job Title block if a safety CSHO or "I," if a health CSHO.

  (b) Item 3 on the certification form shall be filled in by inserting only the word "employer" or "employee" in the space provided. This holds true even when the notice of contest is filed by an attorney for the party contesting the action. An item "4" shall be added where other documents, such as additional notices of contest, are sent to the Commission.

  (c) Use a date stamp with the correct date for each item in the document list under the column headed "Date".

(d) Be sure to have the name and address of the Regional Solicitor or attorney who will handle the case inserted in the box containing the printed words "FOR THE SECRETARY OF LABOR." The Commission notifies this person of the hearing date and other official actions on the case. If this box is not filled in by the Area Director, delay in receipt of such notifications by the appropriate Regional Solicitor or attorney could result.

E. 1. b. (4) Documents Sent to OSHRC. In most cases, the envelope sent to the OSHRC Executive Secretary will contain only four documents--the certification form, the employer's letter contesting OSHA's action, and a copy of the Citation and Notification of Penalty Form (OSHA-2) or of the Notice of Failure to Abate Form (OSHA-2B).

c. Petitions for Modification of Abatement Dates (PMAs).

(1) In accordance with the OSHRC Rules of Procedure the Secretary or duly authorized agent shall have the authority to approve petitions for modification of abatement filed pursuant to 29 CFR §2200.37(b) and (c).

(2) The purpose of this transfer of responsibility is to facilitate the handling and to expedite the processing of PMAs to which neither the Secretary nor any other affected party objects. The Area Director who issued the citation is the authorized agent of the Secretary and shall receive, process, approve, disapprove or otherwise administer the petitions in accordance with 29 CFR §2200.37 and §2200.38, 29 CFR §1903.14a, and D.2. of this chapter. In general, the Area Director shall:

(a) Ensure that the formal requirements of §2200.37(b) and (c) and §1903.14a are met.

(b) Approve or disapprove uncontested PMA's within 15 working days from the date the petition was posted where all affected employees could have notice of the petition.

(c) Forward to the Review Commission within 10 working days after the 15-working-day approval period all petitions objected to by the Area Director or affected employees.

(d) File a response setting forth the reasons for opposing granting of the PMA within 10 working days after receipt of the docketing by the Commission.

E. 2. <u>Transmittal of File to Regional Solicitor</u>.

     a.   <u>Notification of the Regional Solicitor</u>. Under the Commission's Rules of Procedure the Secretary of Labor is required to file a complaint with the Commission within 20 calendar days after the Secretary's receipt of a notice of contest.

     b.   <u>Subpoena</u>. The Commission's rules provide that any person served with a subpoena, whether merely to testify in any Commission hearing or to produce records and testify in such hearing, shall, within 5 days after the serving of the subpoena, move to revoke the subpoena if the person does not intend to comply with the subpoena. These time limitations must be complied with, and expeditious handling of any subpoena served on OSHA employees is necessary. In addition, OSHA personnel may be subpoenaed to participate in nonthird-party OSHA actions. In both types of cases, the Solicitor will move to revoke the subpoena on OSHA personnel. Therefore, when any such subpoena is served on OSHA personnel, the Regional Solicitor shall immediately be notified by telephone.

   3.   <u>Communications with Commission Employees</u>. There shall be no ex parte communication, with respect to the merits of any case not concluded, between the Commission, including any member, officer, employee, or agent of the Commission who is employed in the decisional process, and any of the parties or interveners. Thus, CSHOs, Area Directors, Regional Administrators, or other field personnel shall refrain from any direct or indirect communication relevant to the merits of the case with Administrative Law Judges or any members or employees of the Commission. All inquiries and communications shall be handled through the Regional Solicitor.

   4.   <u>Dealings With Parties While Proceedings Are Pending Before the Commission</u>.

     a.   <u>Clearance with Regional Solicitor</u>. After the notice of contest is filed and the case is within the jurisdiction of the Commission, there shall be no investigations of or conferences with the employer without clearance from the appropriate Regional Solicitor. Such requests shall be referred promptly to the Regional Solicitor for a determination of the advisability, scope and timing of any investigation, and the advisability of and participation in any conference. To the maximum extent possible, there shall be consultation with the Solicitor on questions of this nature so as to insure no procedural or legal improprieties.

E. 4. b.   Inquiries. Once a notice of contest has been filed, all inquiries relating to the general subject matter of the Citation and Notification of Penalty raised by any of the parties of the proceedings, including the employer and affected employees or authorized employee representative, shall be referred promptly to the Regional Solicitor. Similarly, all other inquiries, such as from prospective witnesses, insurance carriers, other Government agencies, attorneys, etc., shall be referred to the Regional Solicitor.

INDEX

OSHA Instruction CPL 2.103
SEP 26 1994
Office of General Industry Compliance Assistance

About Government Institutes

Government Institutes, Inc. was founded in 1973 to provide continuing education and practical information for your professional development. Specializing in environmental, health and safety concerns, we recognize that you face unique challenges presented by the ever-increasing number of new laws and regulations and the rapid evolution of new technologies, methods and markets.

Our information and continuing education efforts include a Videotape Distribution Service, over 200 courses held nation-wide throughout the year, and over 250 publications, making us the world's largest publisher in these areas.

Government Institutes, Inc.
4 Research Place, Suite 200
Rockville, MD 20850
(301) 921-2355

Other related books published by Government Institutes:

Safety Made Easy: A Checklist Approach To OSHA Compliance — W.A. "Tex" Davis, an instructor in the Occupational Health and Safety Department of Texas State Technical College, has joined authors John R. Grubbs, and Sean M. Nelson in providing a new, simpler way of understanding your requirements under the complex maze of OSHA's safety and health regulations. Instead of explaining the labor/OSHA CFR in a chapter format, Davis has created a checklist format to OSHA compliance designed to make your understanding simple and to-the-point. This checklist format is organized alphabetically by topic. Each checklist begins with a brief description of the objectives of the listed items, followed by the required action items and corresponding standards, and where appropriate, training, personal protective equipment, and recordkeeping requirements. *Softcover/190 pages/June '95/$45 ISBN: 0-86587-463-8*

Health & Safety Risk Management: Guide for Designing an Effective Program — This practical guide provides a boilerplate system for any company, large or small, to develop a working health and safety program. A disk may also be added which contains the full text of the book to enable companies to create their own custom programs by just inserting site specific information. *Three-ring binder only/ 310 pages/June '94/$225 ISBN: 0-86587-397-6 Three-ring binder with Microsoft Word files on disk (#4052)/or ASCII disk (#4053)/$250*

Total Quality for Safety and Health Professionals — Total Quality Management is a significant new management concept that is dramatically changing today's business environment. Today, F. David Pierce, a CSP and a CIH, shows you how to apply these concepts - proven successful - to your safety management program to achieve increased productivity, lowered costs, reduced inventories, improved quality, increased profits, and raised employee morale. *Hardcover/230 pages/June '95/$59 ISBN: 0-86587-462-X*

Employment Law Handbook: A Complete Reference for Business —Twelve experts - all highly regarded in their specialty area - have contributed to this handbook, offering unique insights into every aspect of employment law, and offering valuable suggestions as to how to structure your human resources plan. As an employment or labor specialist, or as a manager faced with tough decisions, you will benefit from their experience! *Softcover/330 pages/Dec. '93/$79 ISBN: 0-86587-361-5*